Service Level Management for Enterprise Networks

For a recent listing of titles in the *Artech House Telecommunications Library*, turn to the back of this book.

Service Level Management for Enterprise Networks

Lundy Lewis

Artech House
Boston • London

Library of Congress Cataloging-in-Publication Data
Lewis, Lundy.
    Service level management for enterprise networks / Lundy Lewis.
        p. cm. — (Artech House telecommunications library)
    Includes bibliographical references and index.
    ISBN 1-58053-016-8 (alk. paper)
        1. Business enterprises—Computer networks. 2. Information
technology—Management. I. Title. II. Series.
    HD30.37.L48   1999   99-41775
    658'.0546—dc21   CIP

British Library Cataloguing in Publication Data
Lewis, Lundy
    Service level management for enterprise networks. — (Artech
    House telecommunications library)
    1. Service-level agreements 2. Service-level agreements—
    Data processing
    1. Title
    658'.05

    ISBN 1-58053-016-8

Cover design by Ariana C. Rork

International Standard Book Number: 1-58053-016-8
Library of Congress Catalog Card Number: 99-41775

10 9 8 7 6 5 4 3 2 1

Contents

Preface   ix

    Acknowledgments   xii

1   **Introduction to service level management**   1

    1.1   What is SLM?   2

    1.2   The evolution toward SLM   5

    1.3   The crux of SLM   12

    1.4   Why be interested in SLM?   14

    1.5   Case study: GlaxoWellcome   16

    1.6   Organization of this book   18

    Summary   29

    Exercises and discussion questions   30

    Further studies   31

    Select bibliography   33

2   **Concepts and definitions**   **35**

    2.1   Definitions   36

    2.2   SLM conceptual graph   39

    2.3   Case study: Cabletron Systems and AT&T   44

2.4   A short guide to standards for integrated
management   46

2.5   Comparison with quality of service
management   53

Summary   60

Exercises and discussion questions   61

Further studies   62

Select bibliography   63

| 3 | SLM methodology   **65**

3.1   Essential SLM methodology   66

3.2   An excursion into SE methodologies   74

3.3   Variations on SLM methodology   99

3.4   Case study: Decisys   102

Summary   106

Exercises and discussion questions   106

Further studies   107

Select bibliography   108

| 4 | SLM architecture   **111**

4.1   What is architecture?   112

4.2   Basic SLM architecture   113

4.3   Useful ideas from artificial intelligence,
robotics, and data warehousing   118

4.4   SLM architecture revisited   138

4.5   Evaluating SLM proposals with respect to
architecture   139

4.6   Case study: Deutsche Telekom   142

Summary   151

Exercises and discussion questions   152

Further studies   153

Select bibliography   154

5   Special topics in SLM   157

     5.1   The event correlation problem   158

     5.2   The semantic disparity problem   190

     5.3   The component-to-service mapping problem   196

     5.4   The agent selection problem   207

     5.5   The integration problem   210

     5.6   The scaling problem   212

     5.7   The representation problem   213

     5.8   The complexity problem   215

     5.9   Case study: KLM Airlines   218

     Summary   232

     Exercises and discussion questions   234

     Further studies   234

     Select bibliography   235

6   SLM and electronic commerce   241

     6.1   What is electronic commerce?   242

     6.2   Burdens on suppliers of electronic commerce   243

     6.3   SLM and electronic commerce   245

     6.4   Case study: Windward Consulting Group   246

     Summary   266

     Exercises and discussion questions   267

     Further studies   268

     Select bibliography   270

7   SLM, modern business, and quality of life   271

     7.1   Information systems and modern business   272

     7.2   The SLM connection   275

     7.3   Business trends and challenges   277

     7.4   Why be interested in SLM?   280

     Summary   281

        Exercises and discussion questions   282
        Further studies   282
        Select bibliography   283

Epilogue   **285**

List of acronyms and abbreviations   **287**

About the author   **293**

Index   **295**

Preface

I used to be rather good at working on car engines. As a teenager, I could remove the engine from my '64 Volkswagen, replace a head, put the engine back in place, and be back on the road within three hours—all by myself. I am still rather proud of that and I like to boast about it.

Not anymore. Automobiles are complicated these days. I look under the hood and I don't recognize things. Although I can tell when my Saab is ill, I certainly don't try to fix it myself. I wouldn't know where to start. Plus, I have other things I'd rather be doing: writing books, preparing lectures, and experimenting with networking technology and software. I let the professionals worry about the Saab.

Still, the Saab provides several services without which I couldn't get along. It gets me to work, it gets me to class, it gets me to the airport, it accompanies me on errands, and it takes Dorothy, Lexi, and me on vacations. Let it be out of commission for two days and my life is a mess. Let it be out of commission for a week and—no, let's not go there.

I suspect the reader can identify with these sentiments—if not with automobiles, then with other contraptions.

I hope the reader will pardon the obvious transition to enterprise networks. I can remember an incident in the early 1980s at the University of Georgia when I first saw foreign messages being sent to me on my computer screen. It was scary. The message sender was a professor on the other side of the campus in the Chemistry Department, and I was in the Artificial Intelligence Laboratory.

If I remember correctly, I was running a remote Prolog program that was saturating the professor's VAX machine, on which he was running a precious Fortran program. He was none too pleased about it. His message was for me to stop whatever I was doing. Because he was a professor and I was a graduate student, I acquiesced immediately.

The "enterprise" (it wasn't called that then) provided multiple services for the University. I was running a Prolog application in the interest of AI research. The chemistry professor was running a Fortran application in the interest of chemical research. The enterprise consisted of network devices, computers, and applications running on the computers. Unfortunately, the enterprise could not support the two services at the same time.

The early 1980s were a time when networking technology was a thing for universities. The technology was simple enough so that a reasonably smart person could build a campus network and maintain it, even if the person's education and livelihood were in some other domain.

Not anymore. The deployment and maintenance of enterprise networks are now a thing for professionals. The problems are similar but on a much grander scale. The enterprise still consists of network devices, computers, and software applications. But now there are many more of them, and they are considerably more complex and harder to manage. Further, enterprises are connected with other enterprises via the Internet and third-party backbones, and applications are distributed over all of that. Electronic commerce and distance learning are good examples.

This book is about managing such an enterprise. Our central concept is the notion of a service. Consider that my car mechanic and I share common ground when we talk about the services provided by the Saab and whether the services are in good shape. However, I get lost when he starts talking about catalytic converters and the like, and he gets lost when I start talking about having to get to a conference to present a paper on event correlation.

In the same way, business executives and networking technologists share common ground when they talk about the services provided by the enterprise. But they stray from each other when the executive starts talking about the element of human sympathy in international business or the technologist starts talking about queuing algorithms in switched networks.

Consider the following sequence of definitions:

▶ A *business process* (BP) refers to some way in which a company coordinates and organizes work activities and information to produce a valuable commodity. A typical BP includes several general services in the process, and some of those services depend on the business's enterprise network.

▶ A *service* is a function that the enterprise network provides for the business. We can think of a service as an abstraction over and above the enterprise network. Alternatively, we can think of a service as an epiphenomenon, a phenomenon that arises in virtue of the structure and operation of the enterprise.

▶ An *enterprise network* consists of the following general categories of components: transmission devices, transmission lines among the devices, computer systems, and applications running on the computer systems.

Thus, the concept of service is the intermediate link between business and networking technology. From that concept, we can go upward into the business space and go downward into the technical space. And in this way, we clearly see the connection between the two.

I hope this book will be informative and useful for both business executives and networking technologists. Interestingly, one reviewer of the manuscript argued that I had slighted the concerns of business at the expense of technical detail, while another reviewer thought that I had slighted technical detail at the expense of concerns in business. I made adjustments accordingly; if I'm lucky, I will have gotten the mix just right.

Finally, a third reviewer made the following comment after reading this manuscript:

Lundy: The focus of your book is the enterprise. But do you realize that every service provider in the world is grappling with the same issues that you raise in the book? The managed object may be slightly different; however, the goals and the challenges are identical. With most of the emerging IP service providers, they need to manage applications that are running applications that run on servers over networks (e.g., Web servers, e-mail, and directory applications). Is it too late to generalize to cover service provider environments?

Well, yes, it is too late. The best I can do is quote the reviewer and ask the reader to use imagination.

Acknowledgments

First acknowledgment goes to Dorothy Minior, the first reader of each chapter in the manuscript. Dorothy has a mind for recognizing the difference between clarity and vagueness, logic and sloppiness, and interestingness and triteness. She made sure the chapters were in good shape before I sent them to the official reviewers. She and our dog, Miss Lexi, gave up a lot while I worked on the manuscript, and now I plan to pay them back with considerable interest.

I enlisted the help of several reviewers who have practical knowledge and stake in SLM. Special thanks go to Tom Revak at GlaxoWellcome (United States), Alexandre Passos at NME Ltda (Brazil), and Jim Frey at OSI (United States). These fellows scrutinized each chapter as my manuscript unfolded. Their criticisms were insightful and useful, and I cannot thank them enough for their help. They will see their ideas (and some of their exact words) in the book.

I was fortunate to have had several other reviewers make a pass over the second iteration of the manuscript. All of them provided good comments and criticisms that helped improve the book, and they will see their influence. They are Ed Preston, Mahesh Bhatia, and Audie Hopkins at Cabletron Systems (United States); Minaxi Gupta at the Georgia Institute of Technology (United States); Paulo Rogerio Barreiros d'Almeida Pererra at Instituto de Eng. de Sistemas e Computadores (Portugal); Alex Bordetsky at California State University (United States); Daniel Stevenson at NGI Research (United States); and all of the attendees of my SLM Tutorial at IM '99 in Boston (United States, May 28, 1999).

Two people deserve acknowledgment even though they were not able to review my manuscript: Dr. Pradeep Ray at the University of Western Sydney and Dr. Mani Subramanian at the Georgia Institute of Technology. These fellows were in the midst of writing their own manuscripts, and I was lucky to have the chance to look over their manuscripts before I started mine. They will see their influences.

Katrinka McCollum, general manager at Cabletron Systems, provided wonderful encouragement as I worked on the manuscript. Her support was a blessing.

I thank Bronson Potter (a local scholar, author, inventor, and phi-losopher who lives in my neck of the woods in New Hampshire) for generally hanging out and talking about things. One of the things we talked about was this book.

The people at Artech House were great to work with, and I thank them: Barbara Lovenvirth, Mark Walsh, Tina Kolb, and the anonymous reviewers.

In which we get an intuitive grasp of SLM in preparation for nitty-gritty detail later in the book.

In this chapter:

▶ What is SLM?

▶ The evolution toward SLM

▶ The crux of SLM

▶ Why be interested in SLM?

▶ Case study: Glaxo-Wellcome

▶ Organization of this book

Introduction to service level management

This introductory chapter provides a high-level understanding of service level management (SLM). First, it explains SLM with an ordinary commonsense analogy. Then it provides a brief history of SLM, showing how the evolution of network management has taken us from simple device management in the 1970s to SLM in the late 1990s and into the 2000s.

Next we discuss some reasons why a business would be interested in SLM, after which a true, real-world example of one business's approach toward SLM is presented. Throughout, the book illustrates the discussions and analyses with real-world case studies.

Finally, this chapter gives an overview of the book. The discussion serves as a sort of skeletal map of the book. In addition, the discussion shows that a thorough understanding of SLM will take us into some interesting complementary areas such as artificial intelligence (AI), cognitive science, operations management, and software engineering (SE).

1.1   What is SLM?

SLM refers to the iterative process of (1) identifying business processes, (2) identifying the network services on which the business processes depend, (3) identifying service levels and agents that measure the services, (4) negotiating and articulating a service level agreement (SLA), (5) producing service level reports and comparing them with the SLA, and (6) fine-tuning the enterprise to deliver increasingly better services.

Our definition of SLM is a bit academic. Therefore, let us describe it with a simple analogy. Consider a geographically distributed business. Suppose the business has warehouses and offices in several locations and that there is a need in the business to employ a suite of trucks to transport materials from one place to another (Figure 1.1). Materials might include various things housed in the warehouses, documents sent via interoffice mail, and packages that contain anything from office supplies to computers to desks and bookcases. What is transported is not important. The important thing is that the business needs a suite of trucks to move things from place to place.

Imagine what kinds of functions and personnel are required for a smooth operation of the suite of trucks:

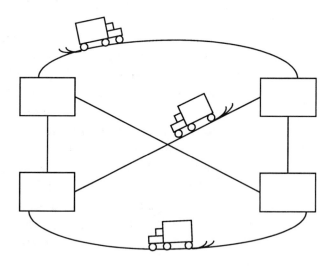

Figure 1.1   A truck operation.

❱ Truck drivers;

❱ Schedulers and route planners;

❱ People to repair and maintain the trucks;

❱ People to load the trucks at a source dock;

❱ People who unload the trucks at a destination dock;

❱ People who monitor the bulk of materials vis-à-vis the number and capacity of the trucks.

We could follow this line of thought further and list other kinds of tasks that make up the general task of material transport. For example, we could take the discussion a level deeper and talk about small trucks and large trucks and partition truck drivers into those who have a license to drive large trucks and those who have a license to drive only small trucks. However, we want only to get the idea of (1) a suite of trucks that supports the transportation needs of a geographically distributed business and (2) an appreciation of the many related tasks that support the smooth operation of the trucks.

Now let us consider the kinds of tasks and services that the trucks support. One, interoffice mail, has been mentioned already. Other services include movement of warehouse goods, office supplies, perhaps classified documents, perhaps cash, or even trash. The type of service will, of course, vary with respect to the charter and the peculiar characteristics of the business. Figure 1.2 illustrates that idea.

The company requires the services listed in Figure 1.2, even though the company's business is not the transportation of materials per se. That observation leads us to another point in our analogy.

Most people in the company do not care too much about the trucks and the tasks required to support the truck operation. Rather, they care about their own jobs and assignments. Those jobs and assignments, however, to some extent depend on the operational efficiency of the trucks.

Consider interoffice mail. How do we measure the goodness of it? And what do we mean by *goodness?* Generally, people in the company who depend on interoffice mail measure it by the efficiency with which it is delivered. As long as nobody in the company complains, the truck

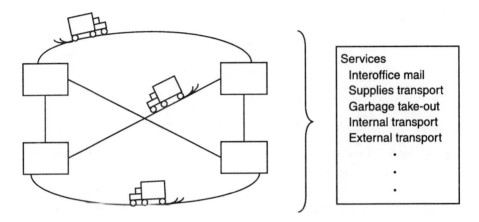

Figure 1.2 A truck operation supporting various services.

drivers and material handlers will go on performing their tasks in their
habitual ways.

Suppose a department began to complain loudly about mail coming
late and in tatters, thereby preventing people from performing their
regular jobs. If late and tattered mail became frequent, the functions of
the department would degrade accordingly. It then would be necessary
for somebody to trace the path of interoffice mail from source to destination to find out where it is being held up or damaged.

Finally, a company might elect to develop and operate a trucking
system internally, or it might elect to hire another company to do it for
them. In either case, it would be important that somebody monitor the
quality of the services provided by the trucking operation. In addition, it
would be important that somebody detect and repair immediate problems. Ideally, someone would extrapolate trends in the company's service
requirements and foresee new service requirements to plan for future
modifications and improvements to the truck operation.

The main points of the trucking analogy with respect to SLM are the
following:

▶ The trucking operation is analogous to the enterprise network of a
business, including the computers and special software applications
that run over the network. In the same way that some services of
a business depend on the suite of trucks, other services depend on
the enterprise network.

▶ The kinds of services supported by the trucking operation are similar to the kinds of services supported by the enterprise network. An obvious example is interoffice electronic mail (e-mail), as opposed to interoffice hard mail. Other examples are numerous, depending on the charter and the purposes of the business in question.

▶ In both domains, the concept of *service* is an abstraction. One would be hard-pressed to point to an entity in the trucking domain that is "interoffice hard mail," just as one would be hard-pressed to point to "interoffice electronic mail" in the enterprise domain. This is a subtle—but important—observation: Services are abstractions that can be decomposed into an aggregate of concrete observable functions.

▶ As in the trucking domain, people who depend on a service supported by the enterprise network look at the well-being of that service differently from the way that the people who provide the service look at it.

Table 1.1 carries the analogy further by listing other similarities between the trucking operation and the enterprise network of a business.

1.2 The evolution toward SLM

This section explains SLM by showing how networking and network management have evolved from simple device management to a state in which SLM is prominent. It is an interesting story, and the evolution toward SLM is logical.

Table 1.1
Similarities Between the Trucking Operation and Enterprise Network

Trucking Operation	Enterprise Network
Roads and trucks	Communications infrastructure
Warehouses	Servers and databases
Goods	Data
People, jobs, tasks	Business processes

Figure 1.3 shows a simple enterprise that consists of two subnets, two routers, two systems on each subnet, and two applications. Application 1 is a local standalone application running on system 1, which is contained in subnet 1. Application 2 is a distributed client/server application running on systems 2, 3, and 4 and thus is supported by subnets 1 and 2, routers 1 and 2, and link 1. Figure 1.3 is used to illustrate the following discussion.

Around the mid-1970s, scientists were just beginning to show that computers could be connected via wires and network devices such as routers. In the beginning, such network devices had to be set up and configured at the source. Network managers had to walk around with dumb terminals, plug them into the devices, and use a special text-based language to manage the devices. We call that process device management (also called element management).

Of course, device management is still going on. Today, however, device management systems are built so that one can set up a device remotely with a graphical user interface (GUI). Any vendor who builds a new router, hub, switch, or other network device also must construct a software application by which the device can be monitored and controlled remotely.

Simple device management carried us up to the late 1980s, at which time it was clear that that way of managing network devices was cumbersome. As network infrastructure grew and became more complex, there was a clear need to centralize the management network functions. Thus were born, around 1990, the so-called network management systems (NMSs). The most notable of this class of management systems

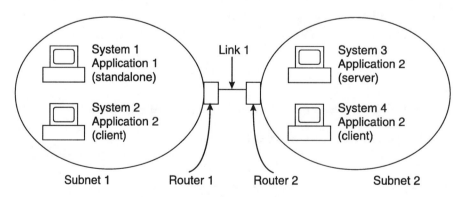

Figure 1.3 A simple enterprise.

were Cabletron Spectrum, HP OpenView, IBM NetView, and SunNet Manager.

About the same time, vendors began to develop products that would allow one to look at the traffic that flows over links in the network. Traffic parameters included volume and type. Early examples of traffic management systems (TMSs) were Silicon Graphic's NetVisualizer and Sun's Etherfind. Today, TMSs incorporate RMON and RMON2 statistics. Good examples are HP's NetMetrix and NDG Software's Programmable RMON II+.

Shortly after the advent of NMSs and TMSs, around 1992, came interest in systems management systems (SMSs). We can think of systems as simple classes of computers, for example, UNIX workstations, PCs, NT workstations and servers, and desktop computers. The idea here is to be able to monitor and control any number of systems remotely with a GUI. Examples of players in this field today are Tivoli's TME, Computer Associates' UniCenter, BMC's Patrol, and Metrix' WinWatch.

It is important to understand the difference between NMSs and SMSs. An NMS has a global view of the structure of the network, including network devices such as routers, repeaters, bridges, hubs, and switches. For example, in Figure 1.3 the NMS would have good representation and control of routers 1 and 2, subnets 1 and 2, and link 1, but little knowledge of the systems other than their presence, location, and whether they were on or off.

Typical pictures in an NMS include topology and logical and geographical views of the network. Typical tasks include autodiscovery, the ability to monitor and control configurations of network devices, and the ability to monitor and produce reports on network performance. An additional important task of an NMS is event correlation, whereby the NMS looks at events issuing from all network devices and infers faults or the general condition of the network as a whole.

SMSs, on the other hand, are concerned primarily with the health of the systems on the network. Typical tasks of SMSs include configuration, performance, and security management. Tasks in performance management include central processing unit (CPU) usage and disk capacity. In Figure 1.3, an SMS would have good monitoring and control capabilities for systems 1 through 4 but little to no knowledge of the network devices and the overall structure of the network topology.

Clearly, NMSs and SMS are complementary and take us a step closer to complete management. However, the NMSs and the SMSs were not integrated in the beginning. Thus, consumers and industry analysts criticized NMS vendors because they were weak in systems management. Likewise, SMS vendors were criticized because they were weak in network management.

One consequence of that state of affairs might have been that the NMS vendors began to spend as much of their resources on systems management as on network management; likewise for the SMS vendors. It became apparent, however, that each of those tasks was huge and required considerable expertise to be performed successfully.

What happened, then, was a thrust for NMS vendors and SMS vendors to integrate their products. By 1993, the concepts of *integration* and *partner programs* and the dictum "No one vendor can provide all the solutions" took firm hold in the industry mindset, and that is a fair description of the state of affairs today.

In 1994, history practically repeats itself with a new concept called applications management systems (AMSs). Vendors began to realize that there is more to the story than network, traffic, and systems management. Although it was a clear achievement to be able to monitor and control a network and the computer systems that reside on that network, vendors began to see that a further achievement would be to monitor and control the business applications that are distributed over networks and computer systems (e.g., client/server applications). Further, vendors began to realize that businesses were more interested in their applications than in the networks and computer systems that supported them.

Some vendors began to study various business applications and to build AMSs that would monitor and control those applications. Examples of business applications are SAP/R3, classic database applications, e-mail, Microsoft Exchange, and Web-based applications. Typical tasks of ASMs are to monitor and report uptime and downtime, performance, response time, and jitter (defined as the variation over response time). Some contemporary examples of AMSs are Platinum's ServerVision, Optimal's Application Insight, and BMC's Patrol. In Figure 1.3, AMSs would have good coverage of applications 1 and 2 but little coverage of the network devices and systems that support them.

So now there are four general categories of players in our story: NMS vendors, TMS vendors, SMS vendors, and AMS vendors. Further, the ideas of integration and partner programs continued to play a large role in complete, end-to-end management solutions.

In 1996, a new term came into being to cover integrated network, traffic, systems, and application management: enterprise management systems (EMSs). Note that nothing really new has been added here, just a new name that denotes integration of the four existing management systems. An example of an EMS is Spectrum/Net Metrix/Patrol/Optimal Insight integrated into a unified whole. Other examples are numerous. Just plug in particular NMS, TMS, SMS, AMS products into the general equation EMS = NMS + TMS + SMS + AMS.

At least conceptually we can see that we have come a long way in just a few years toward total end-to-end management, but the actual implementation and deployment of EMSs are just beginning. In the past few years, there have been quite a number of meetings among consumer and vendor architects to work through the consumer's management requirements, outline an integrated architecture, and proceed to implementation and testing. Similarly, there have been quite a number of meetings among NMS, TMS, SMS, and AMS vendors to establish a business need for integration; if the need is agreed on, there will be more meetings to discuss architecture, design, implementation, marketing agreements, support issues, and so on. There have been successes, partial successes, and downright failures. A subsidiary goal of this book is to put forth good practices toward EMS design and implementation so there are fewer failures.

Now we enter the primary subject of this book: service level management. Imagine a high-level business executive saying, "All this is fine and good. But I need to know how well my business requirements are being met." The executive's concern is with the overall operation of the business, not the technical aspects of network, traffic, system, and application management. Think of the executive's concerns in terms of abstractions over the enterprise network; we can call such abstractions *services*. Thus, the identification, definition, and management of services is *service level management*.

Services in the enterprise space will vary from business to business. Here are some typical examples:

▶ Information transfer among London, New York, and Sydney;

▶ Local information transfer;

- ❯ Web-based price quoting;

- ❯ Web-based electronic commerce;

- ❯ Video conferencing;

- ❯ Voice connections.

It is important to note that we are not speaking of a transition from enterprise management to SLM. Rather, enterprise management is now beginning to include SLM in addition to network, traffic, systems, and application management. Later chapters show that network, traffic, systems, and application management serve as the foundation for SLM.

Clearly, looking at a business's enterprise network in terms of services and SLM raises the bar for EMSs. That way of looking at things is consumer driven and thus holds important challenges for EMS vendors who provide tools for consumers. In general, EMS vendors have to find a way to map elements in a business's enterprise infrastructure (i.e., network devices, systems, and applications) into services and have to be able to measure the well-being of those services. That is the subject of this book.

Concurrent with the publication of this book, people are beginning to think about the next phase of the evolution: business process (BP) management. As a first approximation, we can say that a BP refers to the unique ways in which a company coordinates and organizes work activities, information, and knowledge to produce a valuable commodity. Thus, a BP management system (BPMS) refers to a mechanism to represent, monitor, and control the BP.

A BP will include general services and activities in the process, where some of those services and activities depend on the business's enterprise infrastructure but where other services are unrelated to the enterprise infrastructure. Further, the very idea of a process connotes a temporal dimension. As a simple illustration, refer to the trucking example: A BP might be "Truck all sales receipts to building A by closing time every Monday and enter sales data into the corporate database by closing time Tuesday."

BP management, in the sense that it is described here, is just now drawing interest in the industry. We expect that by the year 2000 the idea of BPMSs will have taken a firm hold in the industry mindset and that vendors will start producing products to support it.

The timeline in Table 1.2 recapitulates the story on the evolution toward SLM since about 1975.

Before leaving this section, let's compare the preceding discussion with some important ongoing work in the standards communities.

The International Telecommunications Union Telecommunications subcommittee (ITU-T) seeks to describe a management infrastructure with standard protocols, interfaces, and architectures. Such a management infrastructure is called a telecommunications management network (TMN). Figure 1.4 shows the management layers of a TMN.

The service management layer is responsible for the management of services provided to customers, including service provisioning, service creation, service contracts, and information about the quality of services. Although the TMN service management layer is not responsible for management of the physical entities per se, the TMN architecture calls for standard interfaces with the lower level layers.

Note the close (but not exact) parallels between the historical account of SLM and the positioning of the TMN service management layer established by ITU-T. On the surface, the obvious difference is the lack of TMN's consideration regarding the play of traffic, systems, and application management toward SLM.

Table 1.2
Timeline of Evolution of SLM

1975	Device management
1990	Network management
1991	Traffic management
1992	System management
1994	Application management
1996	Enterprise management (network + traffic + system + application management)
1998	Enterprise management (network + traffic + system + application + service level management)
2000	Enterprise management (network + traffic + system + application + service level + BP management)

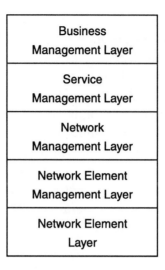

Figure 1.4 TMN management layers.

We must consider, however, that the TMN concept was developed specifically as a service provider tool rather than as an approach to enterprise management. If we simply expand the lower three TMN levels to include traffic, systems, and applications management, the ITU-T standard and our historical perspective of SLM are virtually identical. Both accounts aim toward some way to deal with the complexity in the delivery of services, whether they are in the business enterprise domain or the telecommunications domain.

1.3   The crux of SLM

Later chapters go into considerable detail regarding SLM methods and challenges. There is, however, one particular challenge that most people recognize as the crux of SLM and that should be fully explained at the outset.

Two competing strains are involved in the mapping of the well-being of elements in the enterprise infrastructure into the well-being of services:

   ▶ Parameters that are easy for network specialists to measure do not translate well into parameters that are readily understood by ordinary consumers.

▶ Parameters that are readily understood by consumers are not easy for network specialists to measure.

Let us call this the "semantic disparity problem." The disparity is a result of the way consumers understand their enterprise network and the way network specialists understand it.

For the network specialist, system parameters that are easy to measure include component uptime/downtime and mean time between failure and repair. At the network infrastructure level, parameters that are easy to measure include router and hub statistics such as uptime/ downtime, load, packets lost, and throughput.

On the other hand, the prime parameter that a consumer is interested in, but one that is difficult to measure, is plain user happiness. Measurements of application reliability, response time, and jitter can serve as indices to user happiness, but there is no guarantee.

There is little new in that distinction. It has been many years since Albert Einstein observed that "not everything that can be counted counts and not everything that counts can be counted." The semantic disparity problem is a contemporary instance of Einstein's observation.

Logically, there are three approaches to the semantic disparity problem:

▶ *User-centric approach.* Vendors find some way to measure the parameters of interest to consumers, for example, user happiness or some other measure of well-being with respect to identifiable services.

▶ *Happy-medium approach.* The vendor and the consumer search for service parameters that are easily measurable and also meaningful to the consumer.

▶ *Technocentric approach.* Vendors show consumers how low-level network, traffic, systems, and application parameters translate into higher level parameters that reflect the health of the consumers' services.

Clearly, the preference is the user-centric approach, since consumers are paying for the services. Often, however, the happy-medium approach is sufficient. The simple parameter "port-level availability" is easy for vendors to measure and acceptable to some consumers as an index to service health, both in theory and in practice. Equally often, though, the technocentric approach works. If the vendor is astute enough to understand how high-level service parameters can be composed of lower level parameters and can explain that successfully to consumers, a trusting bond can emerge between the vendor and the consumer.

Although there are no hard and fast rules for choosing the best approach to the semantic disparity problem, it is a step in the right direction for all parties to be aware of the options and the trade-offs of each approach. (Chapter 5 revisits this problem in more detail.)

1.4 Why be interested in SLM?

Now that we have a preliminary understanding of what SLM is and how we got to a state where SLM is prominent, it is a good idea to review reasons for pursuing it.

Among researchers, scientists, and thinkers in general who are nourished by hard problems, SLM is interesting for its own sake. The challenge is difficult and gives us a chance to apply lessons learned in related areas such as AI, SE, operations research, and quality control.

More important is the practical business need for SLM. Consider the following:

- An enterprise network is a commodity for today's global business environment. In short, businesses depend on the unique services that are supported by the enterprise. Thus, it is important for those services to be identified, understood, monitored, and controlled in accordance with some rational procedure.

- Many businesses have a large investment in their networks, systems, and applications. Such an investment is sometimes called the total cost of ownership (TCO) regarding the enterprise. Most businesses, however, have a hard time understanding the extent to which the enterprise contributes to business profit. But if we understand the services provided by the enterprise and the relation between profit and services (i.e., total benefits), then we can calculate the rate of return on investment (ROI). The usual equation for ROI is:

$$ROI = (total\ benefits - TCO)\ /\ total\ initial\ investment$$

▶ SLM methods may help a business unpack the equation to see the utility of expenditures on enterprise components and management tools.

▶ A trend in business is to departmentalize into sales, research, marketing, production, and so on. One such department is the information technology (IT) department. The goal of the IT department is to establish and maintain the enterprise network for the business. Thus, the IT department must understand the service requirements of other departments in the business, for example, the sales, marketing, and research departments.

▶ Some businesses elect to outsource their IT requirements, in which case the IT department is under contract with the business. Other businesses elect to develop their IT department internally. In either case, expectations of services are set and the deliverance of services is monitored, and that calls for some rational procedure by which to do so.

▶ A contributory measure of a business's operational efficiency is the extent to which crucial services provided by the enterprise infrastructure are met. In addition, a good SLM program would identify weaknesses in service provisions, and a better SLM program would suggest ways to rectify the weaknesses.

▶ Finally, a contributory index into a business's value or prospective value is the extent to which the business is operating efficiently. An objective report on a business's service requirements and an evaluation of service provisions provide input into the overall value of the business and thus may influence potential investments into the company.

It is always a good thing for problems to come to the fore that are conceptually and technically challenging and whose solutions would provide clear pay-off for commerce. Such is SLM, whether we are researchers on the cutting edge, vendors who develop SLM products for industry, or executives who oversee business.

1.5 Case study: GlaxoWellcome

This section presents an example of one business's approach toward SLM. The company is GlaxoWellcome (GW), the world's largest pharmaceutical provider.

GW is a classic example of a global business. Its headquarters are in London and Research Triangle Park, North Carolina. Other GW operations are in Italy, Spain, Portugal, and Germany.

Like most global businesses, each GW operation has developed its enterprise infrastructure and management tools more or less in isolation. For example, the London operation uses the NMS HP OpenView to manage its network and a suite of SMSs and AMSs to manage various systems and applications. The North Carolina operation uses Cabletron Spectrum to manage its network and another suite of SMSs and AMSs to manage its systems and applications.

Suffice it to say that there is some overlap in the selection of specific management systems across GW, but there is no agreement among the various operations about which tools should be used. Further, it is all right with GW that there is no such agreement. It would be disruptive to individual operations to start dictating which management systems are to be used, especially when each operation is running smoothly already.

The goal of GW headquarters is to understand and measure the general global communication among individual operations. This is an example of an abstraction over the individual elements (networks, systems, and applications) of the global enterprise. The service is called, simply, intercontinental communication (IC).

Following sound engineering practices, the network architects at GW have mapped out a migration plan toward the IC service requirement. The beauty of the plan is that it is incremental, reasonably but not overly ambitious, nondisruptive, simple, and elegant.

Let's start with the architecture (Figure 1.5). First, observe its simplicity. GW wants to roll up network, systems, and application management information into a consolidated enterprise management system (EMS). The EMS will take as input significant events from the lower level systems, correlate the events, and output enterprise alarms. The alarms, then, will be transformed into trouble tickets and passed automatically to a help desk. In the industry, such an operation is called

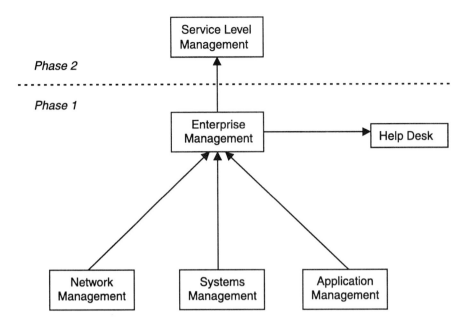

Figure 1.5 The GlaxoWellcome management architecture.

automatic trouble ticket generation (as opposed to user-based trouble ticket generation).

Second, observe that select management information is passed to the SLM system. The SLM system represents the services that GW wants to monitor and control and thus takes as input a select portion of the management information available in the EMS.

Third—and most important—note that the architecture is described without regard to the implementation environment or specific management tools. That is good engineering practice. The overall management system is conceived in an ideal world, purposely thinking that there will be no snags when the architecture is transformed into a design, an implementation, and final deployment. Of course, the GW architects know there will be snags. Nonetheless, they know what the ideal management system is supposed to look like.

Finally, Figure 1.5 shows phases 1 and 2 of the implementation plan. Phase 1 involves the integration of several existing management products selected by GW: Cabletron's Spectrum as the EMS, Remedy's Action Request System as the help desk, and a combination of BMC and Platinum products to cover GW's system and application management requirements. Luckily, all those products are already integrated and available commercially with the exception of Platinum's software, the integration of which is under way at this writing.

Phase 2 is under study. The thinking is that the health of the primary service, IC, can be broken down into two components: the trans-Atlantic link and the Microsoft Exchange server.

If we can successfully measure the health of those components, then we can map the measurements into the overall health of the primary IC service. The SLM products under consideration for the task are ICS's Continuity (Germany), Gecko's SLA Manager (United Kingdom), and Opticom's Executive Information System (United States). Three products are also commercially available and are integrated with the Spectrum EMS.

Further, GW is looking at several products that can measure IC response time from the user's perspective, including Candle's ETE Watch, Jyra's Jyra, Envive's StopWatch, and VitalSign's Vital Agent. One of those monitoring agents likely will become a piece of the consolidated EMS in the center of Figure 1.5.

To sum up, we have given a high-level view of GW's approach toward SLM. Later chapters go into more detail regarding issues and good practices in SLM development and deployment. Some of the issues that will be discussed are (1) good practices toward identification of enterprise services; (2) rationale and patterns for integrating multiple management systems; (3) event correlation over the enterprise; (4) automated monitoring, reasoning, and control regarding enterprise services; and (5) SLM in real-time mode and off-line mode.

1.6 Organization of this book

This book describes state-of-the-science SLM methods for the business enterprise, telcos, and service providers. It covers the following topics:

▶ Concepts and definitions;

▶ SLM methodology;

▶ SLM architecture;

❱ Special topics in SLM;

❱ SLM and electronic commerce;

❱ SLM, modern business, and quality of life.

Chapter 2: Concepts and definitions

Chapter 2 begins with some important concepts and definitions in SLM. The main goal of that chapter is to provide rigor and exactness in the definition of SLM-related terms. Ordinary language can be slippery, and people often miscommunicate because their use of common terms and premises are not agreed upon up front. The purpose of Chapter 2 is to lay common groundwork so that such miscommunication is avoided.

Chapter 2 starts at the highest level, defining a BP and providing examples. The definition of a BP will contain some undefined terms, including the term *service*. Thus, we must proceed to define a service, and so on, until our language and definitions are complete. For example:

❱ *Business processes* are the unique ways in which a company coordinates and organizes work activities, information, and knowledge to produce a valuable commodity. BPs include the use of general services in the process, and some of those services may depend on the business's enterprise infrastructure.

❱ A *service* is an abstraction over components in the enterprise infrastructure, where such components may include any network device, system, application, or communications medium on which the service depends.

In that fashion we proceed to define concepts such as service level parameters, service level agreements, service level reports, and, finally, service level management.

Chapter 3: SLM methodology

Chapter 3 discusses the SLM process, starting with methods for determining whether there is a need for SLM, writing up requirements, conceiving an SLM architecture, and continuing on to design, implementation, testing, and deployment.

Chapter 3 shows that the SLM process is quite similar to SE processes in general; thus, we can borrow and apply some well-known concepts and methods from the SE domain. Looking at SLM as a classic SE task at once makes the whole SLM program tractable and understandable and instills a sense of confidence as we undertake it.

One kind of SLM process was outlined in the case study of Glaxo-Wellcome. In SE jargon, the approach taken by GW is called the *incremental model* of SE. Initially, GW's service requirements included a relatively large number of services. However, GW selected a small number of core services to be worked out in version I space, saving later requirements for version II space, and so on.

Chapter 4: SLM architecture

Chapter 4 takes up SLM architectural considerations. Following good SE practices, the chapter discusses approaches to SLM architecture without regard to design and implementation issues.

Two complementary discussions suffice to show the range of possibilities for a good SLM architecture. Chapter 4 discusses two architectures: (1) an architecture that includes simple monitoring and reporting regarding a collection of services and (2) an architecture that includes monitoring, reasoning, reporting, and control of services. The latter discussion takes us into fully automated SLM in both real-time mode and off-line mode, an ambitious but noble goal. In the industry, it is sometimes referred to as "lights-out" management, that is, management without human intervention.

Figure 1.6 shows the range of possibilities for an SLM architecture. The figure conceptualizes the SLM architecture as a collection of layered control loops and illustrates three layers: a real-time layer, a weekly off-line layer, and a monthly off-line layer. Each layer consists of three processes: interpretation, inferencing, and control.

▶ *Interpretation* (going upward on the left side of the figure) is sometimes called data fusion or data analysis, that is, the processes by which large quantities of data are fused to produce more meaningful but smaller chunks of information.

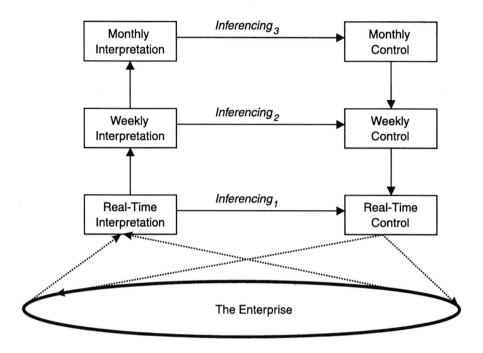

Figure 1.6   Possibilities in an SLM architecture.

▶ *Inferencing* (going left to right at each layer) is the process of making decisions about what actions to take when certain service conditions are violated. On the lowest layer, decisions might include alarm creation, trouble ticket creation, or a recommendation to make an immediate adjustment to a service control parameter to bring the service back in line. On the uppermost monthly level, decision making usually is performed jointly by the consumer and the supplier of a service. That might happen at the end of the month in cases when service reports are unfavorable. Of course, it would be ideal for most decision making and corrections to be done automatically at the lower levels.

▶ *Control* (going downward on the right side) is the process of executing the decision made by the inference layer. A good example is the use of a device configuration product to reconfigure enterprise components to reroute selected categories of traffic.

The illustration in Figure 1.6 is for introductory illustration purposes only. The number of layers does not necessarily have to be three. Nor do all three processes—interpretation, inferencing, and control—have to be in an SLM architecture. The framework, however, provides a good way to situate SLM proposals. In addition, the framework provides the flexibility to tailor specific SLM solutions to meet the needs of a particular business's SLM requirements. For example, we may ask:

 ▶ Are we doing monitoring and reporting only?

 ▶ Are we doing monitoring, interpretation, and reporting?

 ▶ Arc wc doing classic off-line SLM only?

 ▶ Are we doing both real-time and off-line SLM?

 ▶ How much full automation (interpretation, inferencing, and control) do we have? Where?

It is plain to see that when we start talking about automated reasoning without human intervention, we would be well advised to borrow good ideas and methods from communities that specialize in it. Thus, the discussion of SLM architecture in this book borrows and applies some methods from the AI, cognitive science, and robotics communities.

Chapter 5: Special topics in SLM

Chapter 5 discusses some challenging design and implementation problems in SLM:

 ▶ Event correlation;

 ▶ Semantic disparity;

 ▶ Component-to-service mapping;

 ▶ Agent selection;

 ▶ Integration;

 ▶ Scaling;

 ▶ Representation;

 ▶ Complexity.

For each problem, Chapter 5 explores some classic solutions that will get us by, so to speak, and also some more advanced solutions. The fact that there are solutions to each problem is good news. But if we look at the problems more closely, we will see several good candidates for Ph.D. dissertations.

Each challenge is outlined briefly next.

Event correlation

Event correlation is sometimes called root cause analysis. In large enterprises, often a real fault in one area causes several apparent faults in other areas. That phenomenon can cause headaches for administrators who have to sift through a large number of symptoms and figure out the root cause of a misbehaving enterprise.

Section 5.1 defines the event correlation problem and describes several approaches in the industry: rule-based reasoning, model-based reasoning, state transition graphs, codebooks, and case-based reasoning. The strengths and weaknesses of each approach are examined.

Semantic disparity

The semantic disparity problem was discussed in Section 1.3. Simply put, the problem is that the way ordinary users and business overseers look at services provided by the enterprise network is different from the way network specialists look at the same services.

This problem is also called the problem of incommensurabilty: The language and conceptual framework of users and business executives are incommensurate with the language and conceptual framework of network specialists. How can they possibly communicate regarding the goodness and acceptability of services?

Section 1.3 outlined three approaches to the semantic disparity problem:

▶ The user-centric approach;

▶ The happy-medium approach;

▶ The technocentric approach.

Section 5.2 discusses the approaches in more detail and provides examples of each approach, including successes, partial successes, and failures. Most important, we can glean from the discussion some useful guides and heuristics for selecting a particular approach.

Component-to-service mapping

The component-to-service mapping problem is one of finding a function or procedure that takes raw parameters (device, traffic, system, or application parameters) as arguments and provides a value for an inferred, higher level service parameter. That is the challenge, of course, if we elect to take the technocentric approach to the semantic disparity problem.

The simplest solution is the binary function, in which a threshold is set on all low-level element parameters such that any value above the threshold is acceptable. As an example, consider this simple parameter:

$$Up = Yes \text{ or } No$$

That is, we monitor the elements that make up a service as to whether they are running ($Up = Yes$) or not running ($Up = No$).

At first, that sounds fine, but several thinkers have pointed out the dangers of so-called threshold aggregation. For example, if seven elements (say, three network devices, two systems, and two applications) combine to support a service, it is tempting to say that the health of the service is acceptable if each element is up 98% of the time. Note, however, that in the worst case the service could be inoperable 14% of the time, even though we would have to say that the overall metric that had been agreed on had been met.

At the other extreme, AI methods such as fuzzy logic and neural networks have been used to learn the relation between low-level element parameters and consumer attitudes toward the goodness of their services. Those methods show promise, but currently they are, for the most part, in the research stage.

In sum, we can exhibit ingenuity and innovation in coming up with mapping functions for SLM. Look at the task as follows:

$$f(P_1, P_2, \dots, P_n) = S$$

where the Ps are low-level parameters, S is the inferred higher level service parameter, and f is the mapping function. The problem now is to define f. Section 5.3 discusses several ways to define f that are more advanced than the simple binary function.

Agent selection

The problem of agent selection is this: Assume that a network specialist has a good idea of the services a consumer is interested in and thus is in the midst of selecting service parameters and management tools for SLM monitoring. That is just the problem: Which parameters do we measure, and what tools are available for measuring them?

The concept of "monitor agents" is a good way to think about methods for monitoring and measuring low-level service parameters. Monitor agents generally fall into six categories.

▶ *Device agents* have a focused view of the connection nodes in the network infrastructure, for example, bridges, hubs, switches, and routers. Monitored parameters typically include port-level statistics.

▶ *Traffic agents* have a focused view of the traffic that flows over transmission media in the network infrastructure. Examples of monitored parameters include bytes over source-destination pairs and the protocol categories thereof.

▶ *System agents* have a focused view of the systems that live in the enterprise. Typically, these agents reside on the system, read the system log files, and perform system queries to gather statistics. Monitored parameters include CPU usage, disk partition capacities, and login records.

▶ *Application agents* have a focused view of business applications that live in the enterprise. These agents also reside on the system that hosts the application. Some applications provide indices into their own performance levels. Monitored parameters include thread distribution, CPU usage per application, file/disk capacity per application, response time, number of client sessions, and average session length.

▶ *Special-purpose agents* can be built to monitor parameters that are not covered by any of the preceding four categories. A good example is an agent whose purpose is to issue a synthetic query from point A to point B and (optionally) back to point A to measure the reliability and response time of an application. Note that the synthetic query is representative of authentic application queries. An example is an e-mail agent that monitors the response time and jitter of e-mails from one user domain to another.

▶ *Enterprise agents* have a wide-angle view of the enterprise infrastructure, including connection nodes, systems, and applications that live in the enterprise. These agents are also cognizant of relations among the components at various levels of abstraction and are able to reason about events that issue from multiple enterprise components (event correlation or alarm rollup). Monitored parameters that are accessible by enterprise agents are numerous, including router and hub statistics, ATM services, frame relay services, and link bandwidth usage.

Most SLM programs will require a combination of multiple agents of different types. Although there are no hard and fast rules for selecting the right agents for SLM monitoring, Section 5.4 discusses several examples, from which we can glean some heuristics or rules of thumb for selecting the right agents for SLM monitoring.

Integration

Integration is a straightforward extension of the agent selection challenge. Assuming that we are confident regarding our selection of the SLM monitoring agents, we are faced with the problem of integration.

Recall the case study of GlaxoWellcome in Section 1.5. Particularly, refer to Figure 1.5. Several existing management systems (NMSs, SMSs, and AMSs) are in place, but some of them do not talk to each other. Life would be so much easier if they did.

The good news is that they can be made to talk to each other with a little customization in the field. (Fortunately, there is no bad news.) Section 5.5 discusses some classic integration methods so that management systems can pass information back and forth as needed.

Scaling

The scaling problem is this: Consider *end-to-end* SLM, in which we try to cover every possible element that could affect a particular service. Then consider *selective* SLM, in which we cover only a few of all the possible elements that could affect a particular service.

How practical is it to have end-to-end SLM? Certainly, it is feasible if one has the human resources and the monitoring agents at one's disposal. However, it is clearly expensive in terms of deployment and management of SLM.

The distinction between end-to-end SLM and selective SLM is reminiscent of a common distinction in AI between precise, complete solutions and *satisficing* solutions. A satisficing solution to a problem is one that is considerably less expensive than, but not far from, a precise and complete solution. Section 5.6 discusses that distinction in more detail, and points to ways for dealing with the problem of scaling.

Representation

With regard to the representation problem, we have to advance to thinking about the relation between BP management and SLM. (See the preliminary definitions of these two items in the beginning of this section.)

To understand the representation problem, consider a simple analogy, a football team. (We use American football for the analogy, although we probably should use soccer, which is an international sport.) It is straightforward to categorize the players who make up a football team into offensive players, defensive players, running backs, line backers, and so on. It also is straightforward to look at the health of each individual player and to aggregate the individual healths into a single measure of overall health for the football team.

Note, however, that in doing so we do not have a representation of a football *play*. Obviously, the excellent health of a football team does not guarantee that a football play also will be excellent. Something is missing from the representation, namely, the process of playing football and an evaluation thereof.

We have the same problem in representing BPs and their underlying services. It is one thing to pick out the services that support the process, but it is quite another thing to model the process itself. Thus, we distinguish between two modes of representation: component representation and process representation. Tools exist today that can do component representation. Process representation, however, is much harder, albeit quite useful if one can do it. Section 5.7 discusses an advanced method for representing BPs.

Complexity

The complexity challenge is perhaps the hardest challenge of all. Suppose a business has an application that is crucial. A good example is an aircraft scheduling application used at an aircraft control tower. Planes are arriving at the airport from different angles, at different speeds, and under different environmental conditions. Some planes are carrying cargo, and others are carrying passengers. Private jets may be carrying transplant organs, diplomats, network specialists, or business executives.

The scheduling application takes input from several remote sources and gives the controller recommendations about how to schedule the landing sequence of incoming planes. In short, it is a distributed application that depends on the enterprise network.

Clearly, the application must be running at peak performance at all times. Our question is: What affects the performance of this application? Note that we are not asking about the algorithm that does the scheduling. Rather, we are asking about the impact of other components in the infrastructure on the application. That question is difficult for the ordinary human mind to comprehend.

If we can uncover the conditions that lead to good and bad performance of an application, then we are halfway toward ensuring that the application is reliable and trustworthy. Section 5.8 describes some approaches to that challenge.

Chapter 6: SLM and electronic commerce

Chapter 6 considers the special case of SLM with regard to electronic commerce, a fast-growing market. Providers of electronic commerce are beginning to create so-called Web server farms on which industries install their Web sites. Clearly, SLM is important in this arena. Industries have to be assured that their customers can always access their Web sites, that performance will be reasonably good, and that customer transactions are secure. Otherwise, the industry may lose business. Chapter 6 discusses the special requirements for the management of distributed Web server farms and proposes a corresponding architecture.

Chapter 7: SLM, modern business, and quality of life

Finally, Chapter 7 looks at SLM from a broader business and human perspective.

First, we look at the 1970s view of "modern business" and the view of modern business in 2000 and beyond. Both views emphasize the use of information systems in BPs. The latter view, however, includes considerations about the quality of life, and Chapter 7 shows how SLM contributes to that view.

Next, we look at SLM deployment along three dimensions: organization, management, and technology. The technology dimension is the primary focus of this book, although the book would be incomplete if it did not examine the other two dimensions.

The organizational dimension addresses the trend toward the division of businesses into various business units, one of which is usually designated the IT unit. The IT unit is responsible for the information systems, including considerations of SLM, across other functional units in the business. The IT unit may be owned by the company, or it may be outsourced to a third-party contractor. In any case, the IT unit serves as the supplier of services.

The management dimension addresses the monitoring and control of the SLM processes. Administration of SLM can be performed by representatives in the business or outsourced to third-party vendors or consultants. One must beware, however, of the tendency toward skewing the SLM process when SLM administration is under sole control of the supplier or the consumer. It is advisable to have a disinterested party oversee the SLM process.

Summary

This chapter examined SLM from several different angles. First, it considered a commonsense analogy to SLM to obtain an intuitive understanding of *service* and *service level management*. As a first approximation, we defined a service as an abstraction over a set of low-level elements in the enterprise infrastructure, where such elements may include network devices, traffic, systems residing on the network, and applications residing on the systems.

In addition, we showed how SLM evolved from simple device management in the 1970s, to network management in 1990, to traffic management in 1991, to system management in 1992, to application management in 1994, to where SLM is prominent today. The next stage in the evolution is likely to be BP management.

The crux of SLM was identified as being a problem of disparate views of a network: the network specialist's view and the common user's view. That problem affects how each party understands the concept of a service and what constitutes a healthy service.

Next, we provided some reasons why a business would be concerned with SLM. As an example, the case study of GlaxoWellcome discussed a large global business's interest and approach to SLM. GlaxoWellcome's work on SLM will be revisited later in the book.

Finally, we provided an overview of the remainder of the book. Chapter 2 discusses concepts and definitions in SLM. Chapter 3 describes the SLM engineering process, starting with requirements and proceeding to architecture, design, implementation, testing, and deployment. Chapter 4 considers the space of possible SLM architectures. Chapter 5 takes us into SLM technical challenges and solutions. Chapter 6 considers the special case of SLM and electronic commerce, and Chapter 7 considers SLM in relation to the broader picture of modern business and quality of life.

Exercises and discussion questions

1. This chapter used an analogy to describe SLM. Try to think of other domains and tasks that are analogous to SLM. Possible domains might be automobile performance, the health of the human body, or the measurement of human intelligence. Try to pick a domain with which you are familiar. What are the main similarities between the SLM domain and the domain you chose?

2. Consider the technical challenges of SLM described in this chapter in light of your response to exercise 1. Determine whether your analogous domain shares similar challenges and whether there are general solutions or approaches to those challenges that could be carried over to the SLM domain.

3. We have seen the evolution of device management, to network and traffic management, to systems and application management, then to service level management, and finally to BP management. What could possibly come next? Why? Try to give good reasons and arguments for your speculations.

4. This is a research exercise. This chapter hinted that the SLM process shares important similarities with general SE practices. Search the literature and write an essay that answers this question: What SE models exist? Some of the models you might come across in your research are the incremental model, the waterfall model, the spiral model, the clean room model, and the rapid prototyping model. *Note:* This exercise will go a long way toward preparing you for Chapter 3.

Further studies

This chapter assumes general knowledge on the reader's part regarding network technologies such as SNMPv1, SNMPv2, SNMPv3, CMIP, RMON 1 and 2, and others. In academic network management courses and industry tutorials, that information is covered early to provide a foundation for later studies in higher level management problems such as event correlation, capacity planning, enterprise management, and service level management. A very good recent book for such a broad study is Subramanian's *Network Management: Introduction to Practices and Principles.*

Other classic books on the foundations of networking and network management include Terplan's *Communication Networks Management* and Stallings's *SNMP, SNMPv2, and CMIP—The Practical Guide to Network Management Standards.* Our book is a good complementary text for those foundational books in a graduate level network management course or a professional tutorial.

In addition, this chapter assumes a general knowledge of the five classic areas of network management: fault, configuration, accounting, performance, and security management (FCAPS for short). Good books that provide discussions of those areas are Leinwand and Fang's *Network Management: A Practical Perspective*, Lewis's *Managing Computer Networks: A Case-Based Reasoning Approach*, and Minoli's *Enterprise Networking*.

A good book that describes issues and techniques regarding both network and systems management is Hegering and Abeck's *Integrated Network and Systems Management*.

A good starter book that describes ongoing work in the standards community is Aidarous and Plevyak's *Telecommunications Network Management into the 21st Century: Techniques, Standards, Technologies, and Applications*.

The premier conference on the subject of network management is the International Symposium on Integrated Network Management (ISINM), sponsored by the International Federation for Information Processing–Working Group 6.6. The first of those symposia was held in 1989 in Boston and has since grown into a series of biannual symposia. The information published in the proceedings of the symposia is first rate and available in most libraries. See Lazar, Saracco, and Stadler, editors of *Integrated Network Management V*, for sample topics. The motivated reader may want to thumb through the six installments of ISINM up to 1999 and use those books as reference material. Sloman's *Network and Distributed Systems Management* is a compilation of chapters on special topics in network and systems management, some of which were derived from papers presented at ISINM symposia up to 1993.

Another excellent series of proceedings is from the Workshops on Network Management and Control. See Frisch, Malek, and Panwar, editors of *Network Management and Control (Volume 2)*, for a sample of topics.

These days, one can always do a Web search, keying off topics such as "service level management" and "enterprise management." One is likely to find many vendor and industry analyst Web sites that have something to say about SLM, some good and some just so-so. Two very good Web sites are the company ICS Gmbk in Germany (www.ics.de) and Decisys in the United States (www.decisys.com).

Select bibliography

Aidarous, S., and T. Plevyak (eds). *Telecommunications Network Management into the 21st Century: Techniques, Standards, Technologies, and Applications.* IEEE Press, 1994.

Frisch, I., M. Malek, and S. Panwar (eds). *Network Management and Control,* Vol. 2. Plenum Press, 1994.

Hegering, H.-G., and S. Abeck. *Integrated Network and Systems Management.* Reading, MA: Addison-Wesley, 1994.

Lazar, A., R. Saracco, and R. Stadler (eds). *Integrated Network Management V.* London: Chapman & Hall, 1997.

Leinwand, A., and K. Fang. *Network Management: A Practical Perspective.* Reading, MA: Addison-Wesley, 1993.

Lewis, L. *Managing Computer Networks: A Case-Based Reasoning Approach.* Norwood, MA: Artech House, 1995.

Minoli, D. *Enterprise Networking.* Norwood, MA: Artech House, 1993.

Sloman, M. (ed). *Network and Distributed Systems Management.* Reading, MA: Addison-Wesley, 1994.

Stallings, W. *SNMP, SNMPv2, and CMIP—The Practical Guide to Network Management Standards.* Reading, MA: Addison-Wesley, 1993.

Stallings, W. *SNMP, SNMPv2, SNMPv3 and RMON 1 and 2.* 3d ed. Reading, MA: Addison-Wesley, 1999.

Subramanian, M. *Network Management: Introduction to Practices and Principles.* Reading, MA: Addison-Wesley, 2000.

Terplan, K. *Communication Networks Management.* Englewood Cliffs, NJ: Prentice Hall, 1992.

*In which we define
our terms, knowing
that doing so will help
prevent skirmishes
as we begin to exe-
cute SLM programs
in the real world.*

In this chapter:

▶ SLM definitions

▶ An SLM conceptual
graph

▶ Case study: Cable-
tron Systems and
AT&T

▶ A short guide to
standards for inte-
grated management

▶ Comparison with
quality of service
management

Concepts and definitions

The goal of Chapter 1 was to give the reader an intuitive understanding of SLM. We described an analogy to SLM, discussed the evolution of network management toward SLM, and provided a preliminary discussion of the technical issues and challenges in SLM.

This chapter is more analytical. First, it lays out definitions in a systematic fashion and provides examples. Once terms and concepts are firmly in mind, the discussions of SLM methodology, architecture, implementation, and technical challenges in later chapters will flow smoothly.

Next, the chapter illustrates the definitions with a conceptual graph. A conceptual graph is a picture that shows the concepts in a domain and the relations among those concepts. A picture is worth a thousand words, and the picture of the SLM conceptual graph brings the SLM framework together in one place.

Next is a look at a real-world example in light of the SLM definitional framework. The case study describes a collaboration between two companies in the United States: Cabletron Systems as a service provider and AT&T as a consumer. We see how our SLM framework shows the differences between a non-SLM program and an SLM program.

Finally, the SLM framework is compared with ongoing work in the telecom standards communities. The chapter provides a short guide to standards work in integrated management and then looks at recent standards work in quality of service (QoS) management.

2.1 Definitions

This section lays out definitions regarding SLM. Understandably, it is pretty dry going at first. However, the definitional framework livens up when it is complemented with the conceptual graph in Section 2.2 and concrete examples given later in the chapter.

- A *business process* (BP) is the way in which a company coordinates and organizes work activities and information to produce a valuable commodity. A typical BP includes several general services in the process, and some of those services may depend on the business's enterprise network. We are concerned only with those services provided by the enterprise network.

- An *enterprise network* consists of four general categories of components: transmission devices, transmission lines among the devices, computer systems, and applications running on the computer systems.

- A *service* is a function that the enterprise network provides for the business. We may think of a service as an abstraction over and above the enterprise network. Alternatively, we can think of a service as an epiphenomenon, that is, a phenomenon that arises in virtue of the structure and operation of the network.

- A *service parameter* is a variable whose value is an index into the performance of some service provided by an enterprise network.

- A *component parameter* is either (1) a variable whose value is an index into the performance of some component of an enterprise network or (2) a variable whose value controls the performance of some component (e.g., transmission device, transmission line, computer system, or application).

▶ A *component-to-service parameter mapping* is a function that takes as input a collection of component parameter values and provides as output a value for a service parameter.

▶ A *service level* is some value of a service parameter used to negotiate acceptable service qualities.

▶ A *service level agreement* (SLA) is a contract between a supplier and a consumer that identifies (1) services supported by the enterprise network, (2) service parameters for each service, (3) service levels, and (4) liabilities on the part of the supplier and the consumer when service levels are not met.

▶ A *service level report* (SLR) is a document that shows the SLA and a mark showing the actual value of a service parameter over some period of time.

▶ An *agent* is a software entity that monitors, records, and controls values of component parameters. Categories of agents in the SLM domain are network agents, traffic agents, system agents, application agents, special-purpose agents, and enterprise agents.

▶ *Network agents* monitor and control parameters of transmission devices in the network infrastructure, for example, bridges, hubs, switches, and routers. Such parameters typically include port-level statistics.

▶ *Traffic agents* monitor and record traffic that flows over transmission lines in the network infrastructure. Examples of such parameters include bytes over source-destination pairs and protocol categories thereof.

▶ *System agents* monitor and control parameters having to do with computer systems. Typically, these agents reside on the system, read the system log files, and perform system queries to gather statistics. Typical parameters include CPU usage, disk partition capacities, and login records.

▶ *Application agents* monitor and control business applications. These agents also reside on the system that hosts the application. Some applications offer agents that provide indices into their own performance levels. Such parameters include thread distribution, CPU usage per application, login records, file/disk capacity per application, response time, number of client sessions, and average session length. Note that distributed applications may not be manageable by a single application agent. See the description of *enterprise agent,* below.

▶ *Special-purpose agents* monitor and control parameters not covered by any of the preceding types of agents. A good example is an agent whose purpose is to issue a synthetic query from system A to system B and (optionally) back to system A to measure reliability and response time of an application. Note that the synthetic query is representative of authentic application queries. An example is an e-mail agent that monitors success, response time, and jitter of e-mails between user domains.

▶ An *enterprise agent* is an aggregate of any of the other agents described and thus has a wide-angle view of the enterprise infrastructure, including transmission devices, transmission links, systems, and applications that live in the enterprise. Enterprise agents, therefore, are useful for managing distributed applications. These agents are also cognizant of relations among components at various levels of abstraction and are able to reason about events that issue from multiple components, called event correlation or alarm rollup. Clearly, component parameters accessible by enterprise agents are numerous.

▶ *Service level management* (SLM) refers to the process of (1) identifying services, service parameters, service levels, component parameters, and component-to-service parameter mappings; (2) negotiating and articulating an SLA; (3) deploying agents to monitor and control component parameters; (4) producing SLRs; and (5) fine-tuning the enterprise in order to deliver increasingly better services.

The definitions here culminate in a rather cursory but pregnant definition of SLM. Our later chapters on SLM methodology, architecture, and technical challenges will shed further light on the definitional framework.

2.2 SLM conceptual graph

A framework of definitions is a good analytical tool but perhaps not the best means to convey the concepts and terms in SLM. This section complements the definitions given in Section 2.1 with a conceptual graph. Once one gets the gist of the notation of conceptual graphs, the construction and comprehension of the SLM graph are easy.

The notation of the conceptual graph consists of just two kinds of constructs: (1) concepts and (2) relations between concepts. The notational convention is illustrated in Figure 2.1.

Concept 1, at the beginning of the arrow, expresses a subject; concept 2, at the end of the arrow, expresses an object; and the phrase above the arrow expresses some relation that holds between the subject and the object. Thus, the illustration in Figure 2.1 says that business processes are composed of services, not that services are composed of business processes.

We want to express the SLM graph with respect to the principle of Ockham's Razor: Plurality is not to be assumed without necessity. That principle advises that we restrict the concepts and relations in the SLM domain to a minimum, while still being able to express everything we want to say about the domain. The rationale that motivates the Ockham's Razor principle is that an overly rich language promotes confusion, while an overly sparse language promotes a lack of coverage of the domain. We

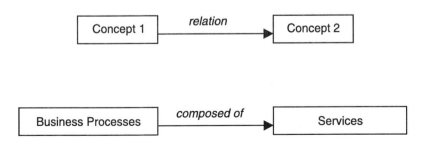

Figure 2.1   Basic notation of the conceptual graph.

want to find exactly the right set of concepts and relations to talk about the SLM domain.

We will build the SLM conceptual graph in increments. The final graph will be somewhat complex, but if we build up to it slowly and conscientiously, it will be palatable. Further, the incremental approach will guide us in asking the right questions at the right time as we begin to plan real-world SLM programs.

To start, examine Figure 2.2. The conceptual graph so far consists of just one relation ("composed of") and three concepts. Also, it is very simple:

▶ Business processes are composed of services.

▶ Services are composed of components.

At the start of a new SLM program, then, we should ask three questions:

1. What business processes are we interested in?

2. What services make up those business processes?

3. What components do the services depend on?

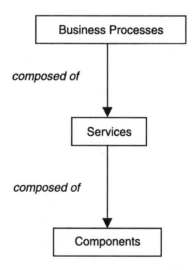

Figure 2.2   First increment of the SLM conceptual graph.

We do not pretend that those are easy questions; rather, they are the right questions once we make the business decision to initiate an SLM program. How we go about answering those and other questions is a concern of SLM methodology, which is taken up in Chapter 3.

We should add that the concept of "services" in Figure 2.2 might well be iterative in the sense that a service is made up of smaller subservices. That typically happens when an external provider offers a particular service that contributes to a company's higher level service requirement. That sort of scenario becomes important in Chapter 3, when we begin talking about SLM methodology.

Now let us take a rather large step to flesh out the SLM conceptual graph further. Take a look at Figure 2.3.

The second increment of the SLM conceptual graph introduces several additional concepts and relations and generates new questions:

4. Once we have identified the services, what are the parameters by which we measure them?

5. Once we have identified the components that make up the services, what are the parameters by which we measure the components? What are the parameters by which we control them? (*Note:* Beware of the distinction between *service* parameters and *component* parameters.)

6. What kinds of agents do we need to monitor and control the values of the component parameters? We can select from device, traffic, system, application, special-purpose, and enterprise agents (assuming such agents are available).

7. How do values of component parameters map into values of service parameters?

Those are difficult questions. In fact, question 7 was highlighted in Chapter 1 as the primary challenge of SLM. The reader should not think that the issues will be sidestepped; we revisit them in painstaking detail later in the book.

Finally, to complete the picture, we need to add the concepts of service levels, SLAs, and SLRs. See Figure 2.4.

SLAs are made up of a list of services and their corresponding service parameters and service levels. The essence of an SLR is a com-

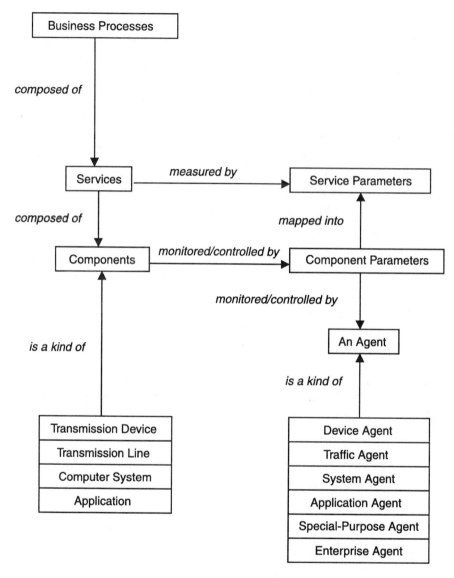

Figure 2.3   Second increment of the SLM conceptual graph.

parison between (1) the actual value of a service parameter over some specified period of time and (2) the mark showing the service level that was agreed on. On the basis of that comparison, one may find reason to reinstrument certain components of the enterprise infrastructure. Thus,

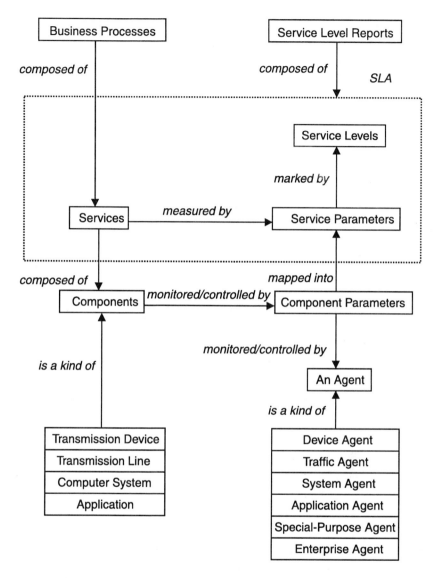

Figure 2.4 Final SLM conceptual graph.

the final increment introduces just one new question. Once we have answered that question, the SLA is practically in hand.

8. How do we go about finding agreeable marks for the service parameters?

Clearly, SLAs also may include other factors: the parties involved in the agreement, the dates during which the SLA is in effect, monies exchanged for services, clauses for reward and punishments, and *ceteris paribus* ("everything else being equal") clauses. In addition, some SLAs in the industry include precise formulas for calculating the values of service level parameters.

2.3 Case study: Cabletron Systems and AT&T

Let us consider a working arrangement between two companies in the United States: Cabletron Systems and AT&T. In 1996, Cabletron Systems made an offer to buy a segment of the AT&T data communications network, which was owned and operated entirely by AT&T, for X dollars and to operate and maintain the network at a monthly charge of Y dollars to AT&T. AT&T accepted the offer.

The monthly charge of Y dollars to AT&T is based on the following calculation. On a monthly basis, Cabletron Systems produces (1) a list of the routers on the network, (2) a list of the ports on each router, and (3) a simple yes/no value as to whether any traffic has passed through the port. Cabletron then multiplies the yes values by Z dollars, and the result is the monthly charge to AT&T.

Of course, Cabletron expects to recover its original X-dollar expenditure over some period of time and then begin to make a profit. Meanwhile, AT&T is relieved of the burden of operating and maintaining the network so it can concentrate on other crucial aspects of its business. Both businesses expect to come out favorably in the long run.

Now, is that what we mean by SLM? Clearly not, but it is a good start in that direction. We have a service provider (Cabletron) and a consumer (AT&T), and we can see some elements of the SLM framework at play. We have a service, that is, provision of the network. We have an identification of the components that make up the service, namely, the total collection of routers (transmission devices). In this case, the service parameter is identical to the component parameter: port usage. Thus, there is a one-to-one mapping of the component parameter into the service parameter. Finally, the software agent that measures that parameter is Cabletron Spectrum, an enterprise agent that monitors and controls the thousands of routers in AT&T's communications network.

The thing conspicuously missing from the Cabletron-AT&T working arrangement is the notion of a service level, an SLA, and an SLR. Although a monthly report shows the number of used ports and the total charge, there is no comparison with some agreed-on service level.

If Cabletron or AT&T were to make a business case for adding an SLM dimension to their working arrangement, there are several possibilities:

▶ Use the same parameter "used ports" and let the service level be this: "The number of used ports will be no higher than 90% of the total." The rationale is that as the percentage of used ports increases, the likelihood of traffic congestion also increases.

▶ Use the component parameter "port load," which measures not just whether *any* traffic has passed through a port but how much has passed through. Let the port load be measured every 10 minutes such that the value represents the percentage of bandwidth utilization during the 10-minute period. Let the service parameter be "free bandwidth" and the service level be that "on average, (free) bandwidth availability will not be less than 30%." The rationale is that when available bandwidth is less than 30%, there is an increased likelihood of congestion and disconnects.

▶ Include a trouble-ticketing service. Let the service parameters be related to the opening of trouble tickets and the repair of network faults "mean time to examine" and "mean time to repair." Let the service level be this: "Trouble tickets opened by users will be examined within 2 hours of submission, and the mean time to repair will not be more than 6 hours."

▶ Establish the service parameter "capacity." Let the service level be twofold: (1) the network will always have enough capacity to accommodate AT&T's crucial business applications, and (2) administrators will receive an alert when capacity is edging toward its limit.

Clearly, Cabletron and AT&T might agree on other services. In fact, such discussions are under way as this book is being written. The goal here, however, is to illustrate the SLM definitional framework on a working example and show how it helps distinguish between SLM programs and non-SLM programs.

2.4 A short guide to standards for integrated management

Our SLM framework shows that services depend on some set of components and that components fall into the following general categories: transmission devices, transmission lines among the devices, computer systems, and applications distributed across all three of those items. If it were only that simple!

Clearly, today's enterprise networks are complex. The current state of any particular business's enterprise more than likely evolved piecemeal and thus includes heterogeneous kinds of components from multiple vendors and various kinds of management techniques. To make matters worse, management techniques vary over countries and over districts within countries.

In a few cases, that state of affairs has resulted in a management nightmare. In other cases, the result is piecemeal management, in which narrowly focused management solutions coexist but do not cooperate with each other. The best case, however, is integrated management, in which management techniques cooperate in a standardized enterprise management scheme.

The ultimate goal of international standards bodies is to provide a uniform framework and methods to correct the current situation. A short guide to standards work in integrated management is presented here. Although the concepts provided are mainly from the telecom point of view, many of them carry over to the data communications network as well.

The five-layer TMN model

A conceptual model of integrated management is shown in Figure 2.5. That model is provided by the ITU-T and is known as a telecommunications management network (TMN). TMN has received general acceptance in both standards communities and industry.

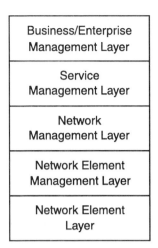

Figure 2.5 TMN model.

In a TMN, management tasks are defined over five layers:

▶ The *business/enterprise management* layer is concerned with the overall management of the telecom carrier business. It covers aspects relating to BPs and strategic business planning. Further, it seeks to capture information to determine whether business objectives and policies are being met.

▶ The *service management* layer is concerned with the management of services provided by a service provider to a customer or other service provider. Examples of such services include billing, order processing, and trouble-ticket handling.

▶ The *network management* layer is concerned with a network with multiple elements. As such, it supports network monitoring and remote configuration. In addition, this layer supports issues such as bandwidth control, performance, quality of service, end-to-end flow control, and network congestion control.

▶ The *network element management* layer is concerned with the management of individual network elements, for example, switches, routers, bridges, and transmission facilities.

▶ The *network element* layer refers to bare elements that are to be managed.

The salient points regarding the TMN model are these:

▶ The TMN is itself a network that monitors and controls another network.

▶ The TMN may be separate from or share facilities with the network it controls.

▶ Every piecemeal management system is meant to be part of an interconnected hierarchy (i.e., the TMN), able to give up its specialized management information to other systems and to ask for specialized management information from other systems.

▶ Each layer in the TMN model is an abstraction over the level beneath it. Tasks at the higher layers are those that need a more abstract view of the network resources; those at the lower levels require a less abstract, more detailed view.

▶ TMN defines standards for interoperability with GUIs such as X-Windows, interoperability with almost obsolete telecommunications network elements, and, more important, interoperability of TMN functions on different layers or within a layer.

▶ The major telecom operators favor the TMN model, and its use in the management of telecommunications systems is well understood. However, it is considered too heavyweight for smaller data networks.

▶ Although the TMN model is meant to include the service and business/enterprise layers, that remains mainly an area of research. Fielded implementations of the model are limited to the network management layer and the element management layer.

Regarding the last point, it should be noted that fielded implementations have accomplished integration beyond the element management and network management layers; however, much of the componentry is custom developed. Although many vendors are supplying point solutions to fit within the TMN hierarchy, none has successfully embraced both the horizontal and vertical dimensions. For example, the company OSI (in the United States) offers pieces that cover a large part of the TMN model, but they are not all compliant regarding interoperability with pieces provided by other vendors.

TINA

Bellcore initiated the telecommunications information networking architecture (TINA) consortium. TINA is designed specifically to tackle the issues in service management. Recall that TMN included this area, although most work on TMN has focused on the myriad challenges involved in the layers beneath the service management layer. Sprint has put together a fielded implementation of a TINA architecture for their ION network.

Figure 2.6 shows a comparison of the TMN model and the TINA model. Particularly, note that the TMN network management layer and element management layer collapse into a general resource layer. That is because TINA is designed specifically with the goals of service management in mind, abstracting away from network details.

A key concept in TINA is the notion of distributed processing and negotiation among a number of agents to realize a service. For example, consider the management of the services offered by a futuristic full-service network (FSN).

The goal of an FSN is to provide a single platform that combines video, voice, and data. An FSN provides a diverse and useful range of services

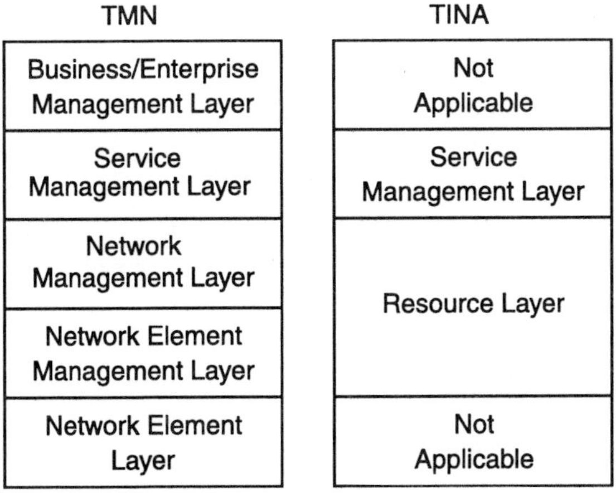

TMN	TINA
Business/Enterprise Management Layer	Not Applicable
Service Management Layer	Service Management Layer
Network Management Layer	Resource Layer
Network Element Management Layer	Resource Layer
Network Element Layer	Not Applicable

Figure 2.6 Comparison of TMN and TINA management layers.

to customers, such as e-mail, video on demand (VoD), interactive television, teleworking, distributed games, video telephony, remote wardens, and access to online public access catalogs. Generally, a range of services that combine video, voice, and data are referred to as *multimedia services*.

A subscriber should be able to choose a service, pay for the service online, then receive the service soon after. The platform takes care of all the technical details of the resources required to deliver and maintain the service during its lifetime. Figure 2.7 shows a collection of agents that might be required to handle such a scenario.

The interface agent is the medium through which the consumer interacts with the rest of the system. The handset and a TV screen might

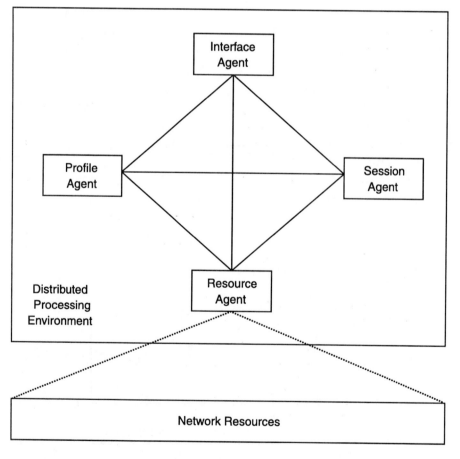

Figure 2.7 Service management via distributed processing.

provide this function. The profile agent holds information about the consumer, such as personal preferences and credit rating. Thus, the selection of some particular service would have to be approved by the profile agent before further processing can happen. Assuming that the service is approved, then the session agent is granted permission to take over. The session agent confers with the resource agent who initializes and maintains resources for the session. The session agent then delivers the service, and is in continuous correspondence with all other agents with the goal of maintaining the service during the session.

This high-level description of the service management scenario sounds appealing, but working out details such as interface protocols and sequencing semantics has proved to be a formidable task. It is tasks such as those that the TINA consortium has undertaken. Understandably, the TINA model has not yet received the general acceptance and popularity of the basic TMN model.

SNMP, CMIP, and CORBA

Refer to the TMN model in Figure 2.5. Clearly, there is a need for communication between each layer of the TMN to realize comprehensive integrated management. If we consider the TINA model (see Figure 2.7), we see that there is a need for communication among agents in one particular layer, the service layer.

The challenge for standards bodies is to specify a standard language by which agents in an integrated management platform communicate, whether they be in a manager-object relationship (i.e., layer N to layer $N - 1$ relationship) or a peer-to-peer relationship (i.e., layer N–to–layer N relationship).

The simple network management protocol (SNMP), produced by the Internet community, is the de facto standard for element management and network management. The great majority of management solutions in the data communications world depend on SNMP to communicate with the network elements.

The structure of SNMP includes two primary components: (1) a structure for organizing information in management information bases (MIBs) and (2) a query protocol to access that information. If a vendor produces a product, whether it is a transmission device or an application, and also includes an Internet-compliant MIB with the product, then the product can be managed by any application that knows the query protocol. The protocol primitives are Get, Set, Get-Next, and Trap.

SNMP is quite simple, which has contributed to its success. However, its simplicity also has resulted in some limitations:

▶ Because the MIB structure is defined at design time, it is difficult to reengineer the MIB in the field if one needs additional data that were not part of the original design.

▶ The MIB structure does not include provisions to represent relations among a collection of managed entities.

▶ SNMP does not support aggregated retrieval facilities, nor does it support facilities to filter retrieved data.

A second potential standard, the common management information protocol (CMIP) developed by OSI, also has two components like SNMP: a management information tree (MIT) and a query protocol to retrieve information from the MIT (Create, Delete, Get, Set, Action, Event-report).

In general, the CMIP model is substantially more complex than SNMP, but there is more you can do with it in terms of management. Thus, there is a trade-off: SNMP is simple to implement and has low overhead in terms of computing resources but lacks expressive power, while CMIP enjoys expressive power but is relatively hard to implement and has high overhead. Neither model provides good facilities to support agents in a distributed processing environment.

A third potential standard, the common object request broker architecture (CORBA), defined by the Object Modeling Group (OMG), provides a computing environment to support multiple collaborating agents. OMG, founded in 1989, is an international nonprofit organization supported by vendors, developers, and users. The CORBA standard comprises:

▶ An interface definition language (IDL) to define the external behavior of agents;

▶ Specifications for building agent requests dynamically;

▶ An interface repository that contains descriptions of all agent interfaces in use in a given system.

The NMF has embraced CORBA for TMN with the emergence of the new so-called SmartTMN. Originally, TMN specified CMIP for all communications throughout the architecture. However, realizing a practical implementation has forced a transition to CORBA for communications above the network management layer. As of this writing, work is ongoing toward definitions of requisite IDLs.

CORBA is expected to be adopted by the TINA consortium. Note that, while it does provide the computing environment for distributed processing, CORBA does not provide a specification for some particular set of agents and agent interfaces required to support a specific task. Those specifications remain to be defined.

Discussion

We have given a short overview of some ongoing work in the standards communities. An important question in the industry today is how the different methods and techniques fit together. One possible scenario is shown in Figure 2.8, in which SNMP is used for element management and network management, and TINA/CORBA is used for service and business management. The gateway between the service layer and the network layer is SNMP based. That scenario, however, is little more than speculation; some working prototypes use SNMP throughout.

2.5 Comparison with quality of service management

Important work is being carried on in the standards bodies regarding quality of service (QoS) management. In fact, it is arguable that if we replace the phrase "service level" with "quality of service" in our discussions of SLM, then (1) nearly all the definitions will have been defined similarly in the field of QoS architectures and (2) the challenges of SLM will be seen as being nearly the same as for QoS. It is incumbent on us, then, to see whether ongoing work in telecom QoS management can help us out with SLM and vice versa.

This section reviews work in progress by the OMG toward a definition of QoS management. That work will result in a set of documents

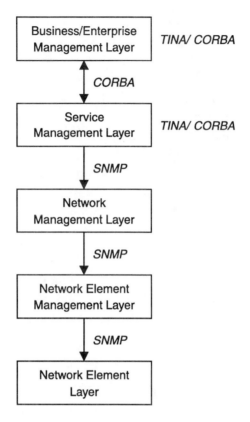

Figure 2.8   A possible configuration for integrated management.

describing QoS standards; currently, it is at the stage of a working draft. As such, the work is available as the *OMG Green Paper on QoS*. The definitive result will be available as the *OMG White Paper on QoS*.

We describe the major concepts that have been developed to date and discuss how they compare to our SLM framework. The goal in the discussion certainly is not to quibble about words and semantics, but to uncover important similarities and differences between QoS management in the telecom standards communities and our SLM framework.

System performance versus system function

OMG view

QoS is a general term that covers system performance as opposed to system function. A system is built to perform some set of functions for its users, for example, holding information. The performance of a system is concerned with properties of functions such as delay, response time, throughput, stability, usability, freshness and accuracy of information, and availability.

Comparison with our view

Clearly, the OMG concepts of function and performance correspond to our concepts of service and service parameter.

User versus engineering viewpoints

OMG view

QoS depends on how a system is being viewed at a given time. A user's QoS requirement may be rapid access to information, but at an engineering level it may be seen as a system function that must itself be performed with a particular QoS, such as a limit on time allowed to complete a transaction negotiation. A delay requirement at the user level may translate into throughput and storage requirements at a more detailed engineering level.

Comparison with our view

The distinction among the user view, the engineering view, and the translations between them corresponds to our distinction among service parameters, component parameters, and service-to-component parameter mappings. From the user's point of view, the hallmark of QoS is that a system (or service or application) does what the user wants, when the user wants it, as the user expected, and as was agreed on.

The question of security

OMG view

Security is seen as a function of a system rather than a quality of some given function. Thus, security functions may have their own set of QoS requirements.

Comparison with our view

Our SLM framework includes very little regarding security. However, we may adopt the OMG stance and say that security is a service, with its own service parameters and service levels.

Static versus adaptive QoS

OMG view

Traditionally, QoS has been something considered during the system design and configuration and engineered statically into the system. Performance goals are established and the systems hardware and software are engineered to meet them based on predictions of the performance of various system elements. Typically, that involves the choices of software and operating system structure, memory size, processing capability, circuit capacity, and so on. But when the system is operational, no further account is taken of QoS.

That approach is not good enough to meet the needs of systems today. The use of multimedia and the extensive use of shared networks for many different and independent traffic streams are leading to the development of systems that can manage QoS dynamically. Such systems can determine the needs, negotiate QoS agreements, and then manage QoS by techniques such as monitoring, admission control, filtering, and application adaptation.

Most systems probably lie somewhere between the static and adaptive ends of the QoS spectrum. The OMG white paper will address the subject of QoS specifications across the whole spectrum, from ways of expressing QoS requirements for static systems to the far more complex mechanisms needed in adaptive systems.

Comparison with our view

The distinction between static and adaptive QoS is perhaps the main point of difference between the OMG framework and our SLM framework, although it is not a crucial difference. Particularly, we are at odds with the idea that once a system is operational, no further account is taken of QoS. In fact, our approach is intended to impose a QoS structure on existing systems for which QoS considerations have been neglected.

This point of difference is not so crucial because the *OMG Green Paper* does indeed plan to address the issues of migration and legacy, for example, how QoS-enabled and current systems interwork.

QoS characteristics and parameters

OMG view

A QoS characteristic represents some aspect of the QoS of a system, service, or resource that can be identified and quantified. It denotes the true underlying state of affairs for the item, as opposed to any measurement or control parameter. QoS characteristics are intended to model the actual, rather than the observed, behavior of the systems that they characterize. For example, the characteristic of transit delay of something between two points is the actual time that occurs between the instants of passage of that thing past those points. The transit delay can never be known exactly although it can be approximated by measurement.

QoS characteristics are distinguished from QoS parameters. A QoS parameter is a value related to QoS that is conveyed between two entities. The information conveyed as a QoS parameter may be of many different kinds, including:

▶ Desired level of a QoS characteristic;

▶ Maximum or minimum level of a characteristic;

▶ Measured value used to convey historical information;

▶ Threshold level;

▶ Warning or signal to take corrective action;

▶ Request for operations on managed objects relation to QoS or the results of such operations.

Comparison with our view

The description of a QoS characteristic suggests Heisenberg's indeterminacy principle, which, roughly put, is the view in physics that we can never know reality as such—we can only know the contents of our minds as we observe reality. Now, we do not have anything like the Heisenberg principle in our SLM framework, nor do we want to take it up (even though we would find ourselves in interesting discussions on philosophy and physics).

We do believe, however, that we are approaching the OMG concepts with our notions of service parameters, service levels, component parameters, and monitoring agents. In particular, a "QoS characteristic" appears to be analogous to our "component parameter," and "QoS parameter" is analogous to our "service parameter." The problem remains, then, to map the former into the latter.

QoS agreements

OMG view

The specification of a QoS agreement relating to a given QoS characteristic includes the following:

- ▶ The particular constraint on the QoS characteristic that is wanted. An agreement can set an operating target, an upper limit, a lower limit, a high threshold, a low threshold, or a combination thereof.

- ▶ What action, if any, is to be taken to ensure the requested QoS, for example, reservation of resources, admission control, or operations of a precedence scheme.

- ▶ Whether the achieved QoS is monitored and, if so, what action is to be taken temporarily or permanently if the requirement cannot be met, for example, initiate an action, abort, adapt in some way, or alert management.

Comparison with our view

Clearly our definition of an SLA is comparable to a QoS agreement.

QoS management functions

OMG view

The term *QoS management function* refers to any function designed to assist in satisfying one or more user QoS requirements. The activities supported by QoS management functions include the following:

> ❱ Establishment of QoS for a set of QoS characteristics;

> ❱ Monitoring of the observed values of QoS parameters;

> ❱ Maintenance of the actual QoS as close as possible to the target QoS;

> ❱ Control of QoS targets;

> ❱ Alerts as a result of some event relating to QoS management.

Comparison with our view

The notion of QoS targets is comparable to our notion of service levels. Chapter 4 examines the functions required to monitor and control service levels in its discussion of SLM architecture.

QoS categories

OMG view

Different user application types will have different requirements for establishing QoS and for controlling and maintaining the actual QoS achieved. For example, the QoS requirements for video streams typically are very different from those for database update transactions.

The different types of user or application requirements, termed *QoS categories,* lead to the choice of particular sets of QoS characteristics to be managed.

Comparison with our view

We have barely mentioned the construct of service templates, which is the counterpart of QoS categories. However, we discuss the development of service templates in more detail in Chapter 3.

QoS management

OMG view

QoS management is the general term for any set of activities performed by a system or communications service to support QoS monitoring, control, and administration.

User requirements, the systems environment, and the systems policies that are in force for the activity drive QoS management activities. User requirements are quantified and expressed as a set of QoS requirements.

Comparison with our view

Our definition of SLM is the same as OMG's definition of QoS management.

Discussion

Generally, we see that there is significant overlap between telecom QoS and our SLM framework. Perhaps the primary difference hinges around the precise, subtle meaning of *service.* We have defined a service as an abstraction or epiphenomenon over network components. Further, we show how a service can be decomposed into some set of components. However, OMG thinks of a service in terms of a collection of agents that collaborate to deliver some function. The service, then, is decomposed into a collection of service agents that communicate with network resources.

Additionally, there is an important difference in the nature of the domains with which QoS and SLM are concerned. Our SLM program is designed primarily for existing enterprise infrastructures, while QoS is designed to be a part of futuristic, next-generation networks. Such networks are sometimes called intelligent networks.

Summary

This chapter presented a set of definitions that lead up to the general concept of *service level management.* The set of definitions was complemented with an SLM conceptual graph, which shows all SLM concepts and relations in a single picture. Next, the chapter examined a working relationship between two companies in the United States: Cabletron Systems and AT&T. We looked at their relationship in light of our SLM framework and showed that the existing state of affairs is not SLM; we also suggested several avenues toward an SLM kind of relationship.

The second part of the chapter switched gears just a bit. We looked at work in the telecom standards community toward integrated management. We discussed two leading integrated management architectures, TMN and TINA, and the management protocols SNMP, CMIP, and CORBA. Finally, we looked at work on QoS and compared it to our SLM framework. We saw that there is fairly large overlap between the QoS framework in the telecom world and the SLM framework in ours.

Exercises and discussion questions

1. Criticize the SLM conceptual graph in Figure 2.4. Are the concepts and relations in Figure 2.4 necessary and sufficient for describing the SLM domain? Redraw the picture with respect to your criticisms.

2. In Section 2.3, *capacity* was suggested as a service parameter. What is capacity? How would you measure it? Are there multiple meanings of capacity? If so, what are they?

3. This is a research assignment. Find out the differences between SNMPv1, SNMPv2, and SNMPv3.

4. This is another research assignment. The OMG seeks to provide CORBA specifications for business objects, electronic commerce, finance, manufacturing, and telecom. Pick one of those topics and describe progress to date.

5. This hands-on exercise assumes that Spectrum or some other appropriate monitoring agent is available. Try to duplicate the existing agreement between Cabletron Systems and AT&T as described in Section 2.3. Produce a list of routers on your network, produce a list of ports on each router, and determine the number of used ports for some small time period. Next, try to implement the first option in Section 2.3 toward SLM. Determine whether the number of used ports is greater or less than 90% of total ports. Produce an SLR that shows the name of the service, the service parameter, the SLA, and a comparison between the SLA and actual measurements.

Further studies

As of this writing there is precious little in the professional literature on SLM. One very good paper is Hegering, Abeck, and Weis's "A Corporate Operation Framework for Network Service Management." For a telecommunications perspective, see the book by Willets and Adams, *Lean Communications Provider: Surviving the Shakeout Through Service Management Excellence.*

In addition, a Web search on "service level management" will turn up white papers from industry vendors and analysts and, if the user is lucky, real SLAs in industry. A very good, albeit quite long, one is "Memorandum of Understanding Concerning Levels of Service Between the Higher Education Funding Councils and the United Kingdom Education and Research Networking Association (UKERNA)." Do a search on UKERNA (if it is still posted).

There is, however, abundant literature on "quality of service management," and we have seen that SLM is closely related to QoS. See Hutchison, Coulson, Campbell, and Blair's paper, "Quality of Service Management in Distributed Systems." Also do Web searches on QoS and "Object Modeling Group" to find standards work on QoS. The personal Web site of Dr. Pradeep Ray (University of Western Sydney, Australia) contains about a dozen papers on QoS. Also see the special issue *Quality of Service in Distributed Systems* (edited by A. Campbell and S. Keshav) in *Computer Communications.* Dr. Campbell is one of the world's foremost authorities on QoS.

The IEEE Communications Society provides concise tutorials on major standards and links to further information from standards organizations, technical committees, and other sources. This service is realized on the Communications Society's Web site (www.comsoc.org).

The discussion of TMN, TINA, SNMP, CMIP, and CORBA was helped by select chapters in Ray's book *Computer Supported Cooperative Work (CSCW)*, Subramanian's book *Network Management: Introduction to Practices and Principles,* and the book edited by Aidarous and Plevyak, *Telecommunications Network Management into the 21st Century.* A Web search on TMN, TINA, SNMP, CMIP, and CORBA will lead to a large body of material. Be forewarned that many people find following telecom standards work to be overwhelming. For more information on OSI's work with the TMN model, visit www.osi.com.

Finally, an excellent opinion paper on future full-service networks, which had considerable influence on the second half of this chapter, is Bhatti and Knight's "On Management of CATV Full Service Networks: A European Perspective."

Select bibliography

Aidarous, A., and T. Plevyak (eds). *Telecommunications Network Management into the 21st Century: Techniques, Standards, Technologies, and Applications.* New York: IEEE Press, 1994.

Bhatti, S., and G. Knight. "On Management of CATV Full Service Networks: A European Perspective." *IEEE Network,* Sept./Oct. 1998.

Campbell, A., and S. Keshav (eds). *Quality of Service in Distributed Systems.* Special issue of *Computer Communications,* Vol. 21, No. 4, Apr. 1997.

Hegering, H.-G., S. Abeck, and R. Weis. "A Corporate Operation Framework for Network Service Management." *IEEE Communications Mag.,* Jan. 1996.

Hutchison, D., G. Coulson, A. Campbell, and G. Blair. "Quality of Service Management in Distributed Systems." In M. Sloman (ed.), *Network and Distributed Systems Management.* Reading, MA: Addison-Wesley, 1994.

Ray, P. *Computer Supported Cooperative Work (CSCW).* Englewood Cliffs, NJ: Prentice-Hall, 1999.

Subramanian, M. *Network Management: Introduction to Practices and Principles.* Reading, MA: Addison-Wesley, 2000.

Willets, K., and E. Adams. *Lean Communications Provider: Surviving the Shakeout Through Service Management Exellence.* New York: McGraw-Hill, 1996.

*In which we learn
some tricks, the use of
which will help ensure
a successful SLM pro-
gram.*

In this chapter:

▶ The essential SLM
  methodology

▶ An excursion into SE
  methodologies

▶ Variations on SLM
  methodology

▶ Case study: Decisys,
  Inc.

SLM methodology

The main goal of Chapter 1 was to provide preliminary understanding of SLM. Chapter 2's main goal was to lay out SLM definitions in a systematic fashion. Now we consider SLM methodology. The goal of this chapter is to learn how to carry out SLM programs in ways that maximize the chance of success.

First, we lay out the essential SLM methodology, which consists of a series of well-defined steps involved in carrying out an SLM program. Next, we take an excursion into SE methodologies and explore the important similarities between SE and SLM. SE has been around for a long time (at least relative to SLM), so it certainly would be to our credit to take advantage of lessons learned from some 25 years of experience in the SE community.

More important, a study of SE methodologies will suggest alternative ways in which we can execute SLM programs. This chapter discusses variations on SLM methodology and potential problems to be on guard against.

Finally, we examine a working example of SLM methodology in the industry at Decisys, Inc., in the United States. We compare the Decisys methodology with the one presented in this book and highlight some important insights into Decisys's experiences in SLM.

3.1   Essential SLM methodology

Our incremental development of the SLM conceptual graph in Chapter 2 gave us a head start toward understanding SLM methodology. Recall the conceptual questions in Chapter 2:

1. What BPs are we interested in?

2. What services make up those BPs?

3. On what components do the services depend?

4. Once we have identified the services, what are the parameters by which we measure them?

5. Once we have identified the components that make up the services, what are the parameters by which we measure the components? What are the parameters by which we control them?

6. What kinds of agents do we need to monitor and control the values of the component parameters? We can select from device, traffic, system, application, special-purpose, and enterprise agents (assuming such agents are available).

7. How do values of component parameters map into values of service parameters?

8. How do we go about finding agreeable marks (i.e., acceptable and nonacceptable service levels) for the service parameters?

We use those conceptual questions to guide us in developing an SLM methodology. Before we proceed, however, let us disclose perhaps the foremost principle of methodology in general:

People who follow a given methodology with exacting precision usually fail and get no fun out of it. But people who adapt a given methodology to their particular circumstances and special talents generally have a higher chance of success and get some fun out of it, to boot.

Besides the assumption that people want to have fun, there are sometimes business and technical reasons for straying from a given methodology; examples are given in Section 3.3. Nonetheless, we want to approach our SLM methodology in a fun sort of spirit, and we advise that similar spirit be cultivated as the methodology is followed in practice.

The SLM methodology presented here is divided into three broad phases.

▶ *Phase 1: Requirements and analysis.* This phase involves performing a preliminary study of a company's BPs and the service requirements. During this phase, technical details are disregarded—the goal is to produce a conceptual model of an ideal SLM system.

▶ *Phase 2: Design, unit testing, and integration testing.* In this phase, the ideal SLM model of phase 1 is carried into the real world in the face of environmental constraints such as network topology and component inventory. In addition, we begin to build nonproduction prototypes in which we test the capabilities of isolated and integrated agents in the SLM system.

▶ *Phase 3: Deployment.* This phase moves the nonproduction SLM system into production. We let the system run for some period of time to establish a baseline and then negotiate an SLA. Finally, the full SLM system is put into operation, producing SLRs in timely fashion and following through per the instructions in the SLA.

For convenience, Figure 3.1 shows the complete methodology on a single page. The reader may want to make a copy to motivate initial SLM discussion or to use as a checklist while working through SLM programs. However, we recapitulate the advice at the beginning of the section: The methodology should be adapted and massaged to fit particular circumstances.

Phase I: Requirements and Analysis

Step 1. The supplier and the consumer work toward a common understanding of the consumer's BPs.

Step 2. The supplier and consumer work toward a common understanding of the enterprise-related services that are required by the BPS.

Step 3. The supplier and the consumer work toward a common understanding of the service parameters and service levels for each service.

Phase II: Design, Unit Testing, and Integration Testing

Step 4. The supplier takes inventory of components: the topology of the network, the kinds of transmission devices and transmission media, the kinds of systems users are using, the kinds of applications that users are using, and existing management processes.

Step 5. The supplier takes first steps toward correlating services and components.

Step 6. The supplier takes first steps toward (1) demarcating component parameters by which to measure and (optionally) control the components and (2) mapping those parameters into service parameters.

Step 7. The supplier takes first steps toward (1) identifying agents to monitor and control components, (2) designing agent integration, and (3) experimenting with nonproduction prototypes.

Phase III: Deployment

Step 8. The supplier moves the overall system into production and a baseline is established to produce the first SLR.

Step 9. The supplier and the consumer review the first SLR, and an SLA is negotiated.

Step 10. Full production proceeds, and SLRs and SLAs are reviewed, followed through, and optionally renegotiated at the end of each time period.

Figure 3.1   Essential SLM methodology.

For example, notice that the consumer disappears in steps 4 through 8 but reappears again in step 9. In many cases, a better approach would be to involve the consumer to some extent during each step of the methodology. The level of consumer involvement may vary as the methodology unfolds, but both parties should be present for the entire process (Section 3.2 explains why).

Phase 1: Requirements and analysis

Step 1. The supplier and the consumer work toward a common understanding of the consumer's BPs.

Discussion   Before taking this first step, it is a good idea for the supplier to do some homework to become familiar with the general language used by the consumer. For example, if a service supplier is scheduled for a first meeting with a health care organization, the supplier would be well advised to study up on health care. Libraries and bookstores have books about the essentials of health care management, and one can find technical papers in journals and conference proceedings (do not refer to articles in popular magazines). Most important, the supplier should lay the materials out at the beginning of the meeting and comment on them. It is surprising the extent to which that promotes goodwill between both parties.

Output   A simple list of terms or phrases that denote each BP and a short description.

Step 2. The supplier and the consumer work toward a common understanding of the enterprise-related services required by the BPs.

Discussion   First, it is important to distinguish between services that depend on the enterprise network and all that is included in it and services that do not depend on the network. Second, it is important to apply simple names to the services. It is a well-known principle in SE that simple, commonsense naming up front will expedite all that follows.

Output   Same as the output of step 1, additionally showing a list of short phrases that reflect each service required by each BP.

Step 3. The supplier and the consumer work toward a common understanding of the service parameters and service levels for each service.

Discussion   For any given service, there will be a multitude of possible service parameters. The supplier needs to know which parameters are most important. Consider a simple analogy. For one package delivery company, perhaps speed of delivery is most important. For another delivery company (one that specializes in fragile cargo), care and caution are more important than speed. Generally, the supplier has to identify the service parameters that have a special relation to the goals of the business.

As in steps 2 and 3, it is important to apply simple names or phrases to the service parameters and service levels. During step 3, it is tempting to be influenced by technical details and environmental dependencies. However, we want to remain nontechnical, almost as if we are doing "armchair philosophy" or living in a perfect world. The rationale is that we know what the ideal SLM structure would look like, although we know it is quite possible we will not be able to achieve it entirely. Nonetheless, we will retain the best features of our ideal SLM structure, and, as a side effect, we will have a useful guide for further research and study.

Output   Same as the output of step 2, now showing a list of short phrases that denote the service parameters and service levels of each service.

Phase 2: Design, unit testing, and integration testing

Step 4. The supplier takes inventory of components: the topology of the network, the kinds of transmission devices and transmission media, the kinds of systems users are using, the kinds of applications users are using, and existing management processes.

Discussion   This step crosses the fine line into technical details. Typically, the person carrying out this step is a network specialist or systems analyst who has a fairly broad knowledge of the variety and functions of network machinery: classic router-based networks, switched networks, virtual local area networks (VLANS), and hybrid switched/router–based networks; transmission media (e.g., Ethernet, token-ring, frame-relay); systems (e.g., UNIX, PCs, NT workstations); and user application types (database, e-mail, client/server, and distributed applications).

We want to produce a high-level, comprehensive picture of the enterprise. We certainly do not want to describe the topology in all its detail, but neither do we want to describe it at such a high level of abstraction that it is meaningless. We want to describe it somewhere in between, and doing that is more an art than a science. There are three useful rules of thumb: (1) convey the structure of the enterprise in a single picture; (2) make use of logical partitions of the enterprise (e.g., departmentwise partitions or geographic partitions); and, most important, (3) test whether a reasonably cognizant person can make sense of the picture within 30 seconds.

Output   A high-level, comprehensive picture of the enterprise.

Step 5. The supplier takes first steps toward correlating services and components.

Discussion   A useful way to carry out this step is simply to lay out one particular service from step 2 next to the high-level topology from step 4. Next, draw arrows from the service to a subset of components in the topology. Beware, however, of the distinction between end-to-end coverage of services and select coverage of services. As an extreme case, consider an e-mail application. If we were to draw arrows from the service "internal communication via e-mail," we would have arrows pointing to practically every component in the topology—all user systems, the mail servers, and all transmission devices and media. That example of end-to-end coverage of the service is not a very viable option. On the other hand, with the selective approach, we might have arrows pointing only to the mail server and the transmission devices.

Output   A combination of the output from steps 2 and 4, namely, a picture showing connections between a service and some subset of components in the comprehensive topology.

Step 6. The supplier takes first steps toward (1) demarcating component parameters by which to measure and (optionally) control the components and (2) mapping those parameters into service parameters.

Discussion   This is perhaps the hardest step of the methodology (the author apologizes for constantly reminding the reader of that fact). The goal is to figure out how to translate low-level component parameters into the service parameters identified in step 3. The easiest way to do that is to declare that some component parameter *is* the service parameter, in which we have a one-to-one mapping between each parameter. A more advanced technique is to devise a function that takes as input a set of component parameters and outputs a composite value of the service parameter, in which we have a many-to-one mapping. Also, note that the input to such a function is likely to be a time series, that is, a table of input values that are measured every 10 minutes. Chapter 5 discusses such techniques in more detail.

Output   Same as the output of step 5 but additionally showing the component parameters for each component and the method that maps those parameters into service parameters.

Step 7. The supplier takes first steps toward (1) identifying agents to monitor and control components, (2) designing agent integration, and (3) experimenting with nonproduction prototypes.

Discussion   Clearly, this step is tightly intertwined with step 6. We want to identify agents (also known as managers), commercial or otherwise, that can indeed monitor the component parameters. Also, we begin to look at the kind of repository that will hold the data collected by the agents and reporting tools for displaying the data. Finally, we have to start thinking about ways to integrate the systems and start experimenting with prototypes. At this stage, then, we begin to build a nonproduction SLM system in which we test the capabilities of isolated and integrated agents in the system.

Output   There are two kinds of output. The first is the same as the output of step 6 but additionally showing the particular agents that monitor and control component parameters, a database into which the data are put, and a reporting tool. Second, we should have a good portion of the overall system actually working. Agents should be monitoring specific enterprise components and populating a database, and we should be able to produce some sample reports.

Phase 3: Deployment

Step 8. The supplier moves the overall system into production, and a baseline is established to produce the first SLR.

Discussion   Finally, we put the system into production in a healthy spirit of hesitation and anxiety. We hope that the bugs have been worked out during testing in phase 2. There should be some procedure by which to shut the SLM system down in case a bug that was missed might affect the normal operation of the enterprise.

Output   Discussion, evaluation, and documentation of what happened.

Step 9. The supplier and the consumer review the first SLR, and an SLA is negotiated.

Discussion   In step 2, the supplier and the consumer made a preliminary pass at identifying the crucial services and service parameters. That understanding might have shifted somewhat by this time. Nonetheless, it is the right time to revisit the initial requirements to see to what extent we are on the mark. Further, it is the right time to negotiate acceptable and unacceptable marks of the service level parameters.

Output   An SLA.

Step 10: Full production proceeds, and SLRs and SLAs are reviewed, followed through, and optionally renegotiated at the end of each time period.

Discussion   The SLA specifies pay-up time, typically on a monthly basis and following usual billing traditions. Thus, monthly SLRs can be considered the same as monthly bills or, in cases in which no monies, rewards, or penalties are specified in the SLA, as simple progress reports.

Output   SLRs every month.

3.2 An excursion into SE methodologies

The practice of SLM is similar to the practice of SE; therefore, we are well advised to take advantage of lessons learned from the SE community. For example, SE typically is defined as follows:

> Software engineering is the establishment and use of sound engineering principles and good management practice and the evolution of SE tools and methods and their use as appropriate to obtain—within known and adequate resource provisions—software that is of high quality in an explicitly defined sense.

This section examines sound engineering principles that have emerged in the SE community and discusses ways in which they can be carried over to our SLM endeavor.

First, we would be well advised to appreciate the magnitude of the risks in SE projects. For example, in the United States in 1995, about $250 billion was spent on approximately 175,000 software projects. However, an estimated $59 billion was spent on cost overruns, and another $81 billion was spent on canceled software projects.

Although we run the risk of overgeneralization, those facts suggest that about 25% of SE projects cost more than the original estimated cost and that another 30% end in failure. If we assume that SE projects and SLM projects share the same risks, then we have about a 45% chance of carrying an SLM program to completion within budget and according to plan.

Why is it that fewer than half of SE projects complete within budget? A recent (1995) study examined opinions of SE project managers in the United States, Finland, and Hong Kong and produced the following consensus of reasons, listed in order of decreasing importance:

1. Lack of top management commitment to the project;
2. Failure to gain user commitment;
3. Misunderstanding the requirements;
4. Lack of adequate user involvement;
5. Failure to manage end user expectations;

6. Changing the scope or objectives;

7. Lack of required knowledge or skills by the project personnel;

8. Lack of frozen requirements;

9. Introduction of new technology;

10. Insufficient or inappropriate staffing;

11. Conflict between user departments.

An examination of the list reveals that the lack of user involvement is perhaps the primary factor that causes an unsuccessful SE program (reasons 2, 4, 5, and 11). A related factor is the lack of a firm establishment and understanding of user requirements (reasons 3, 6, and 8). It is reasonable to group those two factors into one simple, overarching principle:

Get users involved and establish firm requirements up front.

Besides that general admonishment, it is clear that one has to have top management commitment (reason 1) and appropriate staffing and technology (reasons 7, 9, and 10). As an aside, good arguments for getting top management to commit to an SLM program were listed in Section 1.4.

A separate study (1995) of SE projects in England examined the opinions of SE managers who had experienced so-called runaway projects, that is, projects that threatened to spiral out of control. When the managers were asked what they have done in the past to try to gain control of runaway projects, the consensus was this list:

▶ Extend the schedule (85%).

▶ Incorporate better project management procedures (54%).

▶ Add more people (53%).

▶ Add more funds (43%).

▶ Put pressure on suppliers by withholding payment (38%).

▶ Reduce project scope (28%).

▶ Get outside help (27%).

▶ Introduce better development methodologies (25%).

▶ Put pressure on suppliers by threat of litigation (20%).

▶ Introduce new technology (13%).

▶ Abandon the project (9%).

Interestingly, when the same companies were asked what they intend to do in the future to thwart potential runaway projects, the consensus was this list:

▶ Improve project management (86%).

▶ Perform better feasibility studies (84%).

▶ Incorporate more user involvement (68%).

▶ Incorporate more external advice (56%).

The findings in those empirical studies are sobering. We should consider them as potential problems to be on guard against as we undertake SLM programs.

Now, recall that it is the goal of SE methodology to establish and use sound engineering principles to obtain high-quality software within known and adequate resource provisions. In other words, it is the goal of SE methodology to mitigate the risks listed here.

The following sections describe two very popular, complementary SE methodologies that aim to do just that. In addition, they have borne out in practice and have taken a firm root in industry. The two methodologies are the use case methodology and the class-responsibility-collaboration (CRC) methodology. We will see how to take advantage of these methodologies in our SLM programs.

Before we start, there is some encouraging news. The main difference between SE projects and SLM projects is the amount of code that must be written. In SE projects, one typically writes a very large amount of code during the implementation phase. In SLM, however, the code that we write (if any) usually is restricted to integration code, that is, the code that is the glue among agents and management systems in the overall SLM system.

Use case methodology

Figure 3.2 shows the use case methodology in its entirety. We will work through it slowly and conscientiously.

First, assume that we have some sketch of user requirements. It is not so important that the requirements be formal. They could be written on

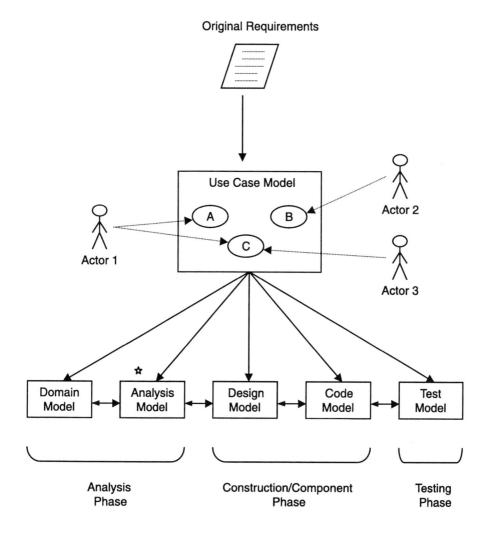

Figure 3.2 Use case methodology.

napkins in a cafeteria. More important is that they express the first sentiments of the users of a software system. They may issue from users themselves, a company's marketing and sales departments, or industry analysts. Generally, it is not advisable that development engineers or computer scientists write up the initial requirements.

Use case model

The first step in the use case methodology is to translate the original requirements into a use case model. The model is a simple picture, much like a child's drawing. It consists of just two kinds of constructs: actors and use cases. An actor represents a user of the system; a use case is a simple phrase that represents how an actor will use the system. The use case model is accompanied by short, clear text that explains the role of each actor and a walkthrough of what happens during each use case.

The most important feature of the use case model is that it is the common understanding between the users of a system and the developers of a system. It is a well-known fact that users and developers employ different languages and have different conceptual interpretations of what a system is supposed to do. That is one of the primary problems in SE project development: What comes out at the end is not what users wanted in the beginning. The prominence of the use case model in this methodology goes a long way toward alleviating that problem.

What would a use case model in our SLM domain look like? Figure 3.3 shows a sample model. There are two actors and five use cases, accompanied by short descriptions. We envision that the supplier and the consumer will use the system the same way; thus, a single actor represents them. A second actor, the overseer, will monitor and maintain the overall system. We do not expect that the supplier or the consumer will be directly involved in this function, although clearly they will have some say-so regarding it.

The SLM use case model in Figure 3.3 is for illustration purposes, but it presents a good starting template. For example, it is reasonable to list the specific services when we describe the View SLR use case. More important, Figure 3.3 shows the right tone and structure: plain, simple, and without regard to technical details.

After the use case model is agreed on, developers proceed to what they do best. Five models are built in increments: the domain, analysis,

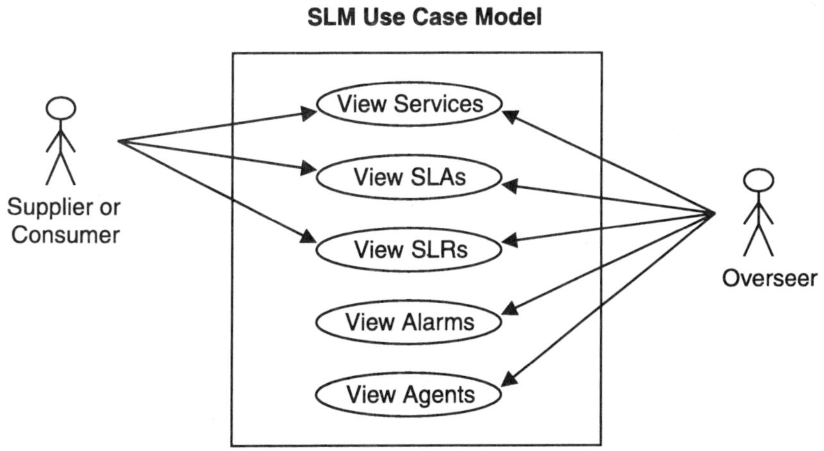

SLM Use Case Model

Supplier or Consumer: Individuals who can view a list of services, view the SLA, and receive SLRs. Billing/accounting is included in the SLR. No modification permissions.

Overseer: Individuals who are general troubleshooters and maintainers of the SLM system. Same viewing rights of supplier and consumer, plus modification permissions such as configuration and setup. Receive SLM related alarms. Has control over agents in the SLM system.

View Services: See a list of services by department.

View SLA: See the SLAs by department.

View SLR: See the SLRs by department.

View Alarms: See SLM-related alarms. Hard alarms notified by pager.

View Agents: Ability to see, monitor, and control agents in the enterprise.

Figure 3.3 Sample SLM use case model.

design, code, and testing models. The use case model and the model that precedes it influence each model. For example, in Figure 3.2 the use case model and the preceding domain model influence the analysis model, as shown by the two arrows leading into the analysis model. Next we show how to construct each of the five models and how each model applies to our SLM domain.

Domain model

The domain model is a preliminary sketch of the objects involved in an SE system, including objects that are in the SE system and the objects outside the system. For the most part, we worked out an SLM domain model when we built our conceptual graph in Chapter 2 (see Figure 2.4). Now we need to draw a boundary that delineates the SLM system from other objects in the domain (Figure 3.4). Enterprise components are considered to be outside the SLM system. The agents that monitor and control those objects, however, are part of the SLM system.

Another object in the SLM domain, which we have not yet talked about, is an alarm. What is an alarm? What kinds of alarms arc involved in SLM systems? And how is the overseer notified when alarms occur?

An alarm is a message to the overseer saying that something is wrong or about to go wrong. Things can go wrong with individual components that make up services. A subtler kind of alarm is when the components seem to be working just fine, but the service is degraded. Thus, there are two general kinds of alarms: component alarms and service alarms.

An event correlation mechanism takes as input a collection of events, scattered in space and time, and maps them into an alarm. That is much what parents do when they observe events happening in their living room while their children are playing, perhaps concluding that a lamp is about to be knocked over.

There are several alarm notification methods in the industry, including paging, phone calls, e-mail, and automatic trouble-ticket generation in the help desk.

We may use the "is a kind of" relationship to show the variety of alarms in an SLM system. Further, we can specify other relations to bring out the general structure of alarm-related objects in the system. For example, Figure 3.5 shows that transmission device alarms, transmission line alarms, system alarms, application alarms, user-generated alarms, and service alarms are a kind of general alarm object. Further, the figure shows possible notification methods and displays a general event correlation mechanism.

Remember that in the domain model we only want to sketch the objects in the domain. We are taking a first step toward identifying and imposing a reasonable structure on SLM objects. For example, the event

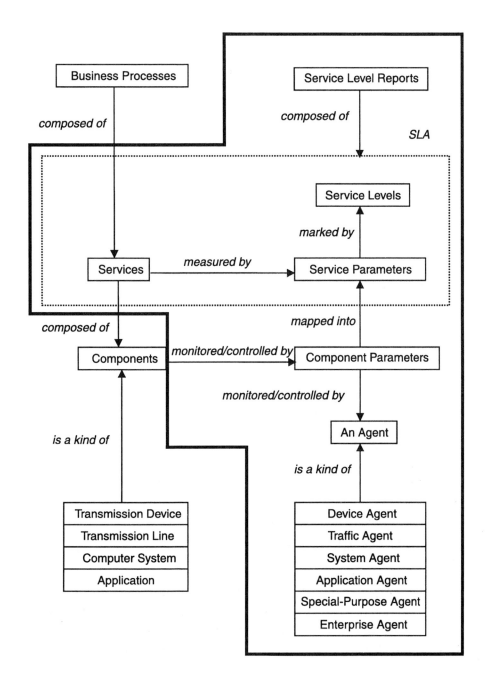

Figure 3.4 SLM domain model with a boundary.

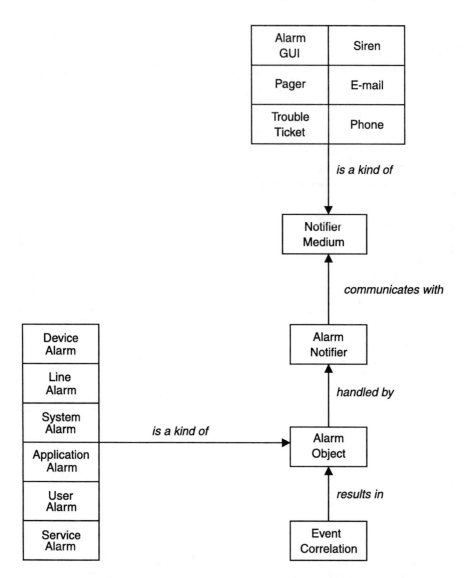

Figure 3.5   Sketch of alarm-related objects in the SLM domain.

correlation mechanism in Figure 3.5 should be considered a black box. It turns out that event correlation is an exciting and challenging task in enterprise management in general, and several approaches are taken in commercial products and research labs. Chapter 5 discusses such approaches.

Analysis model

The analysis model is a step toward the identification of a configuration of objects that will offer each use case in the use case model. We consider that we have access to just three categories of objects: interface objects, entity objects, and control objects. Figure 3.6 shows the symbols that represent the object categories.

▶ Interface objects are the means by which the system connects with the objects outside the domain. The classic example of an interface object is a GUI, in which the external object is just the user at a terminal. Other examples include a command line interface into the system or a database interface.

▶ Entity objects exist for the sole function of holding data. For example, during runtime an entity object may instruct a database interface object to fetch and return a prespecified chunk of data from a database (which is outside the system).

▶ Control objects exist to process data. We can consider control objects as algorithms that take data as input, perform some function over the data, and return a value. For example, a control object may be instructed to perform a trend analysis on data handed to it by an entity object.

A rule of thumb is to never have a particular kind of object performing functions that rightly belong to another kind of object. For example, we do not want to have interface objects processing data or entity objects displaying data. But that is a rule of thumb. Sometimes it makes sense to combine the duties of two objects into a hybrid object.

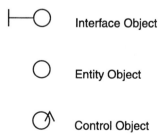

Figure 3.6   The three categories of objects.

As an example, let us consider the View SLR use case in Figure 3.3 and show how collaboration among objects can offer it. Figure 3.7 shows a first approximation of an analysis model for the View SLR use case. The accompanying walkthrough text is self-explanatory.

Now, to complete a comprehensive analysis model for our SLM system, we would sketch models for each use case and then converge

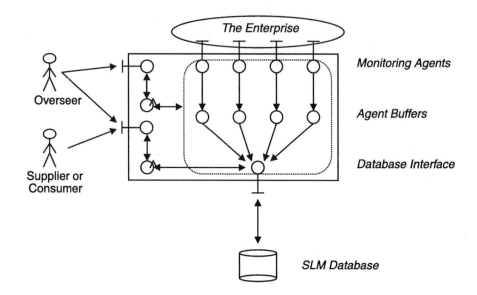

Walkthrough of the View SLR Use Case

The overseer, the supplier, and the consumer use the same GUI interface object to get SLRs. On demand, the GUI object sends an instruction to a control object, which in turn sends an instruction to a database interface object to fetch the data. The control object receives the data, performs a component-to-service mapping function, and sends the result back to the interface object for display.

The overseer uses a separate interface mechanism to configure the agents that monitor components in the enterprise. Agents may include network device, network line, systems, or application agents. Each agent has a temporary buffer to store data. At prespecified intervals, the buffer is flushed and sent to an SLM database via the database interface object.

The viewing of SLRs presupposes that the SLM database has been populated.

Figure 3.7   Analysis model for the View SLR use case.

them. We would see that some objects would participate in all use cases, whereas other objects might contribute to only one use case.

We can imagine what additional objects would be required for the View Alarms use case. Suppose we have a requirement for both service alarms and component alarms, in which the supplier/consumer needs to know only about potential service alarms but the overseer needs to know about all of them. Further suppose that our event correlation mechanism is a simple threshold function.

For service alarms, it seems obvious to incorporate the threshold function into the existing control object. We could build a timer in the control object that periodically fetches component data, computes the component-to-service mapping, and runs the result through the threshold function. Thus, the control object would act like a daemon that runs in the background in addition to its normal function of preparing data for SLRs on demand by the user.

For component alarms, one option is to insert a control object incorporating a threshold function between each monitoring agent and corresponding buffer agent. An alternative option is to incorporate threshold functions into the existing interface agents, in which case we have hybrid monitoring agents. For the sake of simplicity, it is reasonable to choose the latter.

The time is ripe at this juncture to explain the arrow that connects the domain model to the analysis model in Figure 3.2. Note that the arrow is a double-barbed arrow, with the right barb being bigger than the left barb. That means we want to carry as much information as possible from the domain model to the analysis model, although we allow some backtracking and perhaps a minimal amount of back-and-forth play between them.

For example, development of the analysis model might uncover some objects that were overlooked in the domain model, or it might cause us to rethink the boundary that separates SLM objects from non-SLM objects. It is entirely acceptable if the analysis model influences the domain model (if we have done a good job developing the latter, the influence will be minimal). That point holds true as we move through each model in the methodology, as Figure 3.2 shows.

Design model

Recall that the analysis model in Figure 3.7 was conceived without regard to implementation constraints or available tools. We consider the analysis model an approximation of the ideal system and assume that we will have all the tools we need when we go about implementing it.

When we construct the design model, we face reality. We try to find tools, commercial or otherwise, that fit the structure of the analysis model. Also, we have to pay attention to performance and usability issues. There are two reasons for taking that approach:

- In making the move from the analysis model to the design model, we retain the maximal amount of structure of the ideal system.

- What is left over serves as (1) a guide for further research and development for academic/industry labs and (2) a guide for product requirements to be considered by commercial vendors and integrators.

Chapter 6 provides a review of commercial tools and shows how they match up to common objects in an ideal SLM analysis model for electronic commerce. Here, we give just one example of a transition to a design model.

Consider the discussion of component alarms in the analysis model. We know that alarms require an event correlation mechanism. Our question was where that event correlation mechanism should reside. Should it be a separate control object apart from the monitoring agent, or should it be a part of the monitoring agent itself, in which case we have a hybrid control-interface object? We chose the latter for the sake of simplicity and elegance.

It turns out that some commercial monitoring agents have event correlation mechanisms built into them and some do not. Everything else being equal, then, our analysis model advises that we select commercial monitoring agents that have it. Otherwise, we would have to revisit our analysis model and adjust it accordingly.

Next, consider that the analysis model in Figure 3.7 shows four classes of agents that monitor the enterprise: device, traffic, system, and application agents. Now, many large enterprises have thousands of transmission devices and ten times more systems and applications.

That would require tens of thousands of agents to monitor and control the enterprise components. Clearly, a collection of tens of thousands of agents likely would result in a performance strain, a deployment strain, a configuration strain, and a budget strain. It is entirely possible that such a design model would have a negative impact on the normal functions of the enterprise.

A study of commercial products will show that the popular enterprise management platforms integrate multiple agents in a single system, and a few of them have an event correlation mechanism built into them. Those are so-called enterprise agents (see the definition in Section 2.1).

Good candidates for enterprise agents in the industry are Cabletron Spectrum and HP OpenView. Those enterprise agents do a good job at network, systems and application management, but they generally are lacking in traffic management. For example, Cabletron Spectrum is integrated with popular systems and application management products such as BMC Patrol, Platinum ServerVision, Metrix WinWatch, and Tivoli TME.

A candidate traffic-monitoring agent is the Programmable RMON II+ agent from NDG Software. A feature of NDG's traffic agent is that the overseer can write traffic management routines in programming languages such as Perl and then download them to the traffic-monitoring agent. Because monitoring and calculation are done close to the source, there is less effect on the normal functions of the enterprise.

A candidate SLM application is the commercial product Continuity, developed by ICS GmbH in Germany. A feature of Continuity is that it is integrated with Cabletron Spectrum, which in turn is integrated with the products mentioned in the preceding paragraphs. Further, Continuity contains template SLAs and SLRs for common services and standard algorithms for rolling up component parameters into service parameters.

Finally, a candidate for the SLM database is the Spectrum Data Warehouse. That product is designed to interface with enterprise management systems and allow further development of off-line management applications such as accounting, capacity planning, and data mining.

A potential design model for the View SLR use case is shown in Figure 3.8. It is important to note how the picture represents a transformation of the analysis model in Figure 3.7. One should be able to understand clearly the mapping of the objects of the analysis model in Figure 3.7 to

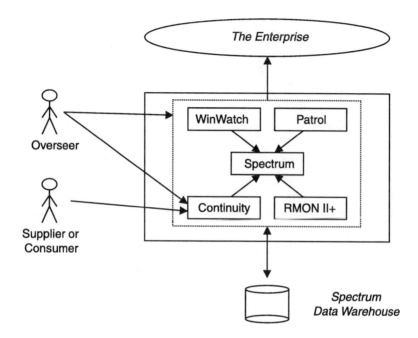

Walkthrough of the View SLR Use Case

The overseer, the supplier, and the consumer use Continuity to generate SLRs. On demand, Continuity performs a component-to-service mapping function, using data in the Spectrum Data Warehouse, which has been populated by Spectrum, WinWatch, Patrol, and RMON II+ monitoring agents. Integrated event correlation and alarming are performed by the Spectrum enterprise monitoring agent.

The overseer uses a common integrated interface to configure the agents that monitor components in the enterprise, configure SLAs and SLRs, and manage alarm notifications.

The viewing of SLRs presupposes that the Data Warehouse has been populated.

Figure 3.8   Design model for the View SLR use case.

the commercial products in Figure 3.8. Further, note how the walk-through text in Figure 3.7 is transformed into the walkthrough text in Figure 3.8. Further note how we pay tribute to the original View SLR use case in Figure 3.3. If that is not clear, then something is wrong with our engineering.

The author emphasizes that this is a candidate design model. Our advice to the reader is to be able to articulate a use case model (and, if possible, a complementary analysis model) and convey it to companies who offer SLM solutions. Then simply compare how their proposed design models match up with the requirements.

Code model

In the use case methodology, the code model is simply the code itself. The code in SLM systems usually involves integration code. Fortunately, we do not have to worry too much about developing software systems from scratch. Rather, we have to worry about how existing software systems work with each other to realize the design model and, by implication, the analysis model and use case model.

It turns out that our candidate design model is already fully integrated except for the NDG RMON II+ agent. Thus, we have a partially complete SLM system, but we have more integration work to do. One option, of course, is to do the work. Another option is to study what else is available in the traffic agent category. For example, RMON agents from the company Netscout are integrated with Spectrum. A study would examine Netscout's and NDG Software's RMON agents to compare their features and determine whether it is worth the extra work to take advantage of any special features of NDG's RMON agents.

Test model

Like all other models, the test model is derived from the use case model and the model that precedes it, in this case, the code model. Unit testing will have been done during construction of the code model. It is important to note that the walkthrough of each use case in the original use case model is mapped directly to the test model. That ensures that what the system does at the end of the methodology is what was planned for at the beginning of the methodology.

Other important considerations

Refer to the big picture of the use case methodology in Figure 3.2. Thus far, we have covered everything in the methodology except two important features: (1) when to start developing prototype GUIs and (2) the component portion of the construction/component phase.

It is common in SE projects to develop a prototype GUI straight from the original user requirements. Sometimes the GUI is constructed with pencil and paper. Other times it is constructed quickly with GUI development environments such as Visual Basic. In either case, what is expected to take place behind the GUI is imagined or staged.

The use case methodology advises that that is a bad idea. What happens is that the prototype GUI begins to influence what the system is supposed to do, when we really want our understanding of what the system is supposed to do to influence the prototype GUI. Further, we begin to understand what the system is supposed to do somewhere around the beginning of the analysis model. The star in Figure 3.2 shows that. The general advice of the use case methodology is simply to think about the system for a while before constructing prototype GUIs, that is, somewhere in the analysis phase.

We can apply that piece of advice to building any kind of prototype, for example, prototyping an integration of two management tools. In general, building a prototype before the requirements are understood can have significant impact on the total cost of product development. Additional costs and frustration often arise in having to reengineer a product to meet the requirements that were not understood the first time around.

The component portion of the construction/component phase is one of the strongest features of the methodology. The word *component* in this context should not be confused with the way we have been using it thus far. A component in a software system is something that is specifically designed so it can be reused in other software systems. Components can be blocks of code, entire applications, or designs.

The benefits of component reuse are saved time and money as we develop other software systems, although we will have spent a little extra time and money in developing them the first time around. However, one should beware that the reuse of components should be considered against a clear understanding of a set of requirements. Otherwise, one runs the risk of premature prototyping all over again, where the component becomes in effect the prototype. That is an easy trap to fall into.

The examples that follow illustrate the benefits of component reuse.

Clearly, we use the component reuse strategy at the software system level when we build SLM systems. It would be a terrible thing if we were to develop all the monitoring agents and reporting systems from scratch each time we approached a new SLM system.

At the documentation level, we can certainly reuse the models and designs that we built in our first SLM system. That includes the final SLM user manuals.

As a final example, consider the alarm notifier object in Figure 3.5. An embodiment of an alarm notifier object is included in the Spectrum Enterprise Management System. In fact, it is called AlarmNotifier. It originally was developed to notify administrators of alarms via e-mail. The engineers, however, took the extra time to provide hooks so it can be integrated with other software systems, for example, trouble-ticketing systems, paging systems, and building blueprint systems. Fortunately, their extra effort has paid off. The AlarmNotifier has been used over and over again to integrate Spectrum with other enterprise management systems. Many of those integrations are done at customer sites in less than a day.

Things to keep in mind about use case methodology

The overall picture in Figure 3.2 is a good summary of the use case methodology. We should keep in mind several main points of the methodology as we develop SLM programs:

- The ultimate goal is to avoid the kinds of risks discussed at the beginning of this section, the main one being to avoid a divergence from user specifications during development.

- The use case model is a good way to express a common understanding among users and developers of SLM systems.

- Developers and integrators should be careful not to stray from the original use case model. As each incremental model is developed, the use case model should be observed.

▶ During the analysis phase, in which we build the domain and analysis models, implementation details should not bog down developers and integrators.

▶ Although it is not likely that we will have a perfect transformation from the analysis model to the design model, we retain as much of the ideal structure as possible and have a good indication of further work to do.

▶ One should not be premature in building prototype GUIs.

▶ It is worth the extra time and effort to build components that can be reused.

▶ Components should be reused as often as possible, including documents, designs, existing software systems, and code.

CRC methodology

The class-responsibility-collaboration (CRC) methodology has become popular in both industry software development departments and university classes that teach object-oriented (OO) methodology. Typically, CRC is combined with an OO language such as Smalltalk, C++, or Java when system designs are implemented.

There is a fair amount of overlap in the use case methodology and the CRC methodology. For example, the term *use case* means the same as the CRC term *scenario*. The domain model and the analysis model are much the same as the CRC exploratory phase and analysis phase.

Comparisons of terminology, semantics, and notation, however, invite confusion. Instead, we will examine useful features in CRC that have not been covered but that complement the use case methodology. First, we look at the major concepts from which CRC gets its name: classes, responsibilities, and collaborations. Next, we look at the main thing that makes CRC methodology so appealing to software engineers: class cards. Finally, we look at the notion of subsystems.

Classes, responsibilities, and collaborations

A class is an abstraction over a collection of objects, and it is related to the objects by the "is a kind of" relation. We have uncovered several kinds of classes and objects in our SLM domain already. Figure 3.4 shows an agent class and several objects that are kinds of agents. Figure 3.5 shows an alarm object class and a notifier-medium class.

A class hierarchy shows how various classes are related to each other. For example, we can imagine extending the system alarm class in Figure 3.5 to show that UNIX OS Alarms and NT Alarms are kinds of system alarms. Further, we could decompose UNIX OS Alarms into thread alarms, login alarms, and CPU alarms, which also might be kinds of NT Alarms. Finally, we should note that classes do not have to be part of a class hierarchy. An example is the alarm notifier in Figure 3.5. We might say that such objects are in a class by themselves.

The responsibilities of a class include (1) actions that the class performs and (2) information that the class holds. We considered generic responsibilities of three classes when we discussed interface, entity, and control objects. The CRC methodology is more specific.

Consider the alarm object class in Figure 3.5. One responsibility of an alarm object is to hold information about itself. Such information might include alarm ID, type of alarm, time of the alarm, severity of the alarm, the agent that issued the alarm, the component to which the alarm applies, the location of the component, the IP address, the MAC address, the underlying events that caused the alarm, the probable cause of the alarm, and perhaps a recommendation about how to deal with the alarm.

A second responsibility of an alarm object is to give up information about itself when asked or to vanish when told to do so. Otherwise, an alarm object does not seem to do much else. It is a rather static object and a good example of a classic entity object.

On the other hand, consider the alarm notifier in Figure 3.5. The information that the alarm notifier class holds is its process ID, its state (e.g., idle or nonidle), CPU usage, and the agents to which it is connected. Its primary responsibilities, however, are to receive alarm objects and to forward them to some notifier medium. Thus, the alarm notifier object is mainly a control object.

In the CRC methodology, class responsibilities that are publicly accessible are called *contracts*. A contract is akin to a promise to perform a function when asked to do so by another class. The main contract of the alarm object is to give up information on demand. The main contract of the alarm notifier is to forward alarm information.

A class may also have private responsibilities. Private responsibilities are functions that the class must perform to do its job, but the functions are not considered publicly available. For example, a private responsibility of the alarm notifier is to check for duplicated alarms before forwarding them to a notifier medium.

Collaboration is a communication between one object and a set of other objects so that the one object can fulfill its responsibilities. The contract "forward alarm information" of the alarm notifier in Figure 3.5, for example, requires a collaboration of the alarm object and the notifier medium.

A collaboration graph shows how objects work together to fulfill their individual contracts and private responsibilities. There is a special notation in CRC methodology for constructing such graphs, but to avoid confusion we will not go into that notation here. Suffice it to say that the illustration in Figure 3.5 is a close approximation to the notation.

Class cards

The three simple concepts of class, responsibility, and collaboration go a long way toward demarcating and describing the objects that will make up a software system. A class card in the CRC methodology is the tangible embodiment of those concepts, which is why it is so appealing to software engineers.

A generic class card is shown in Figure 3.9. The contents of the card should be self-explanatory in light of the preceding discussion. A few additional points are these: A concrete class is one that can be instantiated, while an abstract class exists for taxonomic purposes only and thus cannot be directly instantiated.

The signature of a contract or responsibility is the method of invocation. For example, if a contract of the trouble-ticket class in Figure 3.5 is "enter alarm information in the database," then the signature of the contract might be:

> *EnterTT (id type time severity source component location IP MAC event_list)*

Imagine a stack of class cards (or a loose-leaf notebook, in which the cards are sheets of paper) that embodies a complete software system. The stack is completely cross-referenced such that for any object, an engineer can navigate to (1) its position in a class hierarchy graph, (2) its position in a collaborations graph, and (3) its collaborator classes.

```
┌─────────────────────────────────────────────────────────────┐
│  Class: class name                        Concrete or Abstract│
│  Superclasses: class names                                    │
│  Subclasses: class names                                      │
│  Hierarchy Graph:    page number                              │
│  Collaboration Graph:    page number                          │
│  Description:   a short paragraph                             │
│                                                               │
│  List of Contracts                                            │
│          Contract Name:    name                              │
│          Description:   a short paragraph                    │
│          Signature:  method of invocation                    │
│          Collaborations:   list of classes and their contracts│
│                                                               │
│  List of Private Responsibilities                             │
│          Responsibility Name:   name                         │
│          Description:   a short paragraph                    │
│          Signature:  method of invocation                    │
│          Collaborations:   list of classes and their contracts│
└─────────────────────────────────────────────────────────────┘
```

Figure 3.9   Generic class card in CRC methodology.

Software engineers like to lay out the class cards on the floor or a large table to be able to see the overall structure of their design and look for ways to improve it. That technique has proved to be an expedient way to come up with good software designs.

Because each class contains a signature, engineers are a step closer to implementing the system. Once the design is satisfactory, they proceed to distribute class implementation assignments over a group of developers, fairly well assured that it will all come together in the end.

The CRC methodology was conceived as an aid to developing software systems from scratch. In our SLM domain, however, we know that we are mainly integrating existing commercial software systems. Nonetheless, we can take advantage of the techniques in CRC.

Figure 3.10 shows a class card for a trouble-ticketing system. We have taken the liberty of adding another category to the card: possible embodiments. An embodiment of a class is simply a candidate commercial (or homegrown) product that meets the requirements of the class proper.

Class: *Trouble-ticket system*   Concrete
Superclasses: *Notifier medium*
Subclasses: *None*
Hierarchy Graph: *Figure 3.5*
Collaboration Graph: *Figure 3.5*
Description: *This class is the central repository for lodging
problems, notifying administrators and maintenance personnel,
escalating problems, imposing a workflow on the problem repair
process, and maintaining repair statistics.*

List of Contracts
    1.   Receive alarms from management systems
        Description: *accept alarms from peer management
        systems and transform them into trouble tickets*
        Signature: *EnterTT (X, Y, Z, ...)*
        Collaborations: *Alarm Notifier*

         ⋮

List of private responsibilities
    1.   Set timers on trouble tickets
        Description: *a timer on a trouble ticket will cause
        notification to the administrator if the ticket has not been
        acknowledged*
        Signature: *configured through administrator GUI*
        Collaborations: *none*

         ⋮

Possible Embodiments
    1.   Clarify
        Reference:
        Contact:

    2.   Peregrine ServiceCenter
        Reference:
        Contact:

    3.   Remedy Action Request System
        Reference:
        Contact:

         ⋮

Figure 3.10   Class card for a trouble-ticketing system.

Note that this is a clear way to initiate the step from an analysis model to a design model.

It is a straightforward extension to model services with service graphs and service cards and add them to the stack. A service graph shows a service, service parameters, and underlying components (see Figure 3.4). A service card is the same as the generic class card in Figure 3.9 but will include an additional slot for a component-to-service mapping function. Both service graphs and service cards are cross-referenced with other cards in the stack in the usual way.

Subsystems

A final concept in CRC is the notion of subsystems. Imagine how a large number of class cards on a large table might become complex. Engineers often impose a subsystem structure on such an arrangement to reduce complexity. A subsystem is simply a logical grouping of objects that combine to perform some identifiable function.

Figure 3.11 shows a monitoring subsystem, a reporting subsystem, an alarm management subsystem, and a user interface subsystem, all of which work together to offer the overall SLM system. Note that the monitoring subsystem collaborates with the other three.

The subsystem structure should make sense at a glance. Its picture is usually the first card in the stack, followed by subsystem cards that reference other classes and collaboration graphs that uncover the real complexity of each subsystem. For example, the monitoring subsystem would reference a picture like Figure 3.7, in which the rounded rectangle encloses the objects that compose the monitoring subsystem.

Things to keep in mind about the CRC methodology

The most useful features of CRC are (1) the explicit concepts of classes, responsibilities, and collaborations and (2) the development of a software design by the use of class hierarchy graphs, collaboration graphs, class cards, and subsystems.

It is easy to see how we can use similar concepts and methods when we design SLM systems. The first thing we have to do is jump up a level of abstraction. That is, we have to consider objects as existing software systems such as monitoring systems, event correlation systems, reporting

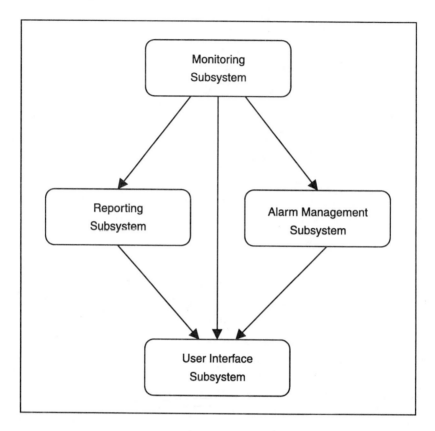

Figure 3.11   Subsystems of an SLM system.

systems, and trouble-ticketing systems. It is straightforward, then, to think in terms of how those software systems collaborate with each other to provide a function that none of the systems can provide in isolation.

We extended the notion of a class card to include a category for possible embodiments, in which an embodiment is an off-the-shelf system that incorporates the responsibilities and collaboration requirements of the class specification. Think of that extension as a good way to take the step from an ideal analysis model to a design model, typically the hardest part of software development.

Also, we added the constructs of service graphs and service cards, a natural extension of the CRC methodology to the SLM domain.

3.3 Variations on SLM methodology

We have taken a rather long excursion through two popular SE methodologies and shown how we can apply their concepts and techniques to SLM system development. Now it is time to revisit the essential SLM methodology in Section 3.1.

First, note that the SLM methodology in Figure 3.1 is presented as a stepwise procedure starting at step 1, almost like a recipe. Real software development is not exactly like that, and neither is SLM development.

Authors of SE texts call that the classical linear approach to system development. They typically set up that approach as a sort of straw man and proceed to uncover its weaknesses and develop more robust methodologies accordingly.

It is important to note that the classical linear approach is never completely abandoned. It is improved on or varied in certain ways to fit special circumstances. In similar spirit, we discuss variations on the essential SLM methodology next.

Backtracking

Backtracking from one step to a prior step in the SLM methodology is expected. Sometimes we have to backtrack just one step, other times we might have to backtrack further, perhaps all the way to step 1. Recall that backtracking and back-and-forth interplay among models are realistic features of the use case methodology. The same principle applies to the steps in our essential SLM methodology, although we want to minimize it as much as possible.

At the same time, however, we do not want to fall into the mode of *thrashing.* Thrashing occurs when architects find themselves retracing steps back and forth repeatedly. Unfortunately, thrashing is a common malady in both SE development and SLM development.

There is no common cure for thrashing except getting a good night's rest and starting fresh in the morning. A complementary cure is to read, reread, and then read more about SLM. The select bibliographies in this book provide a good starting point.

The "version 1 space, version 2 space, and so on" strategy

A common problem to guard against in software development and SLM development is trying to do too much. Biting off more than you can chew can manifest itself in the thrashing phenomenon.

For example, step 1 calls for a list of BPs. Step 2 calls for a list of services that support each BP. It is a good idea at that juncture to prioritize both the BPs and the services before proceeding to step 3. A rule of thumb is to put those services that support the core business high on the list. If the business depends heavily on electronic commerce via the Web, then the Web services should be high on the list.

After the list is prioritized, it is wise to pick, say, the top three and call that "version 1 space." The next group of services will be called "version 2 space," and so on.

Version 2 space can be initiated at some point when version 1 appears relatively stable and well understood. To be safe, step 1 of version 2 would start around step 9 of version 1 space. When following this strategy, it is possible to have several versions going on at the same time, each at different levels of maturity. Note that this strategy requires careful management practices.

To see a variation on the version N space strategy, the reader should reread the case study of GlaxoWellcome's approach to SLM in Chapter 1. Note that SLM is not included in version 1 space at all. Version 1 consists of putting the enterprise management system in place, which includes integrated network, systems, and applications management. Then, when the enterprise management system is stable, GW will implement SLM in version 2 space. That strategy makes sense, because SLM depends on network, systems, and application management.

The "starting in the middle" strategy

In some cases, it makes sense to start at step 4, work backward to step 1, and then forward again. Step 4 calls for an inventory of the enterprise components, including the number and kinds of transmission devices, transmission media, computer systems, and applications.

When would such a strategy make sense? A common business partnership in the industry is the following: A network service provider buys a network that is owned by a company and then takes responsibility for maintaining the network and periodically extending it to support additional services required by the company.

Clearly, the first question that the provider and the consumer want to answer is how much the network is worth, and that calls for taking inventory. Then the provider and the consumer follow the essential methodology in the usual way. A good example of the strategy is described in the case study of Cabletron Systems and AT&T in Chapter 2.

The "starting at the bottom" strategy

In the industry, the strategy of starting at the bottom is sometimes called a "point solution" strategy. It also can be called the "everything is already worked out for a specific domain of interest" strategy.

Consider a hardware vendor that specializes in one kind of enterprise component, for example, a frame relay network. Frame relay networks are used for real-time voice transfer and are useful adjuncts to other network technologies that can support real-time voice transfer such as ATM.

A vendor that produces frame relay gear will have defined the services that its networks can offer and will have developed management agents that can monitor and report the performance of frame relay service parameters. For example, there are two modes of frame delivery in frame relay networks: (1) committed delivery with guaranteed bandwidth and (2) bursty, best-effort delivery. The frame relay forum has begun work to standardize the service parameters latency, jitter, and error rate over frame delivery.

Thus, the point solution already will have been worked out until step 8 for a specific set of services for a specific enterprise component. The main difference, however, is that step 8 specifies that the *overall* SLM management system is put into place. It is likely that the overall BPs and services will have to be revisited, starting at step 1 and proceeding forward in the usual way. Point solutions usually turn out to be one piece of the overall SLM solution.

Some people argue that the combination of point solutions can turn into a "cat-herding" adventure. The analogy has strong effect if one imagines what it would be like to herd a group of cats. The gist is that without clear communication and a commonly understood direction, the combination of point solutions will generate more problems and inefficiencies than it solves. The strategy in the next section will help alleviate the cat-herding problem.

The "starting at the top and the bottom at the same time" strategy

This strategy combines a top-down business-oriented strategy with a bottom-up technology-driven analysis. All businesses that depend on enterprise networks have some set of enterprise management procedures in place. It makes good sense to (1) establish business goals and supporting services (phase 1), (2) study the current management practices (phase 3), and then (3) determine what can be done to modify current management practices to meet business goals.

This approach assumes that some kind of production management system is in place (compare step 8). Typically, there exists some set of agents and management methods in place for fault management, configuration management, and performance management. Often, however, the agents are not integrated into a comprehensive enterprise management system. Management methods may be somewhat piecemeal and scattered.

Nonetheless, the existing management methods may constitute a part of the overall SLM management system. It remains, then, to ascertain the business goals and supporting services, as advised in steps 1 and 2, and to see to what extent the existing management methods can support them. We can describe this strategy colloquially as the "what can we do with what we have, and what do we need to do more?" strategy.

3.4   Case study: Decisys

Because SLM is just coming into prominence, there is not enough evidence to compare and evaluate alternative methodologies and subtle methodological variations in a scientific fashion. The best we can do is to take instruction from general SE methodologies, which was done in the preceding sections.

The importance of SLM development and management methods will vary from business to business. A global business that depends heavily on global electronic commerce will be quite interested in SLM, while a business that depends only on local interoffice mail will be less interested.

The following communication from a colleague in Brazil sums up the current state-of-affairs in SLM nicely.

> One of my intentions with the small business is dealing with the problem of SLM management methods. We wish to establish some kind of neutral methodology that can be applied to small/medium networks in order to have a better-managed environment. We know that larger networks deserve much attention; yet companies that have a well-established methodology sometimes run into trouble when deploying their products and techniques, particularly when it comes to integrating their products with products supplied by others and home-developed products. However, I remain optimistic even though I often have to adopt someone else's particular methodology when I deploy their products.

It seems clear that our colleague in Brazil is dealing with vendors who practice the point solution strategy. The obvious challenges are how to establish a vendor-neutral methodology and how to go about integrating point solutions into an overall SLM framework. Hopefully, the instruction in this book will help.

(As an aside, our colleague in Brazil not only is looking for a vendor-neutral methodology but is also a reviewer of this manuscript. Perhaps by the end of the book we will be able to revisit his predicament and see what progress has been made.)

The remainder of this section examines a vendor-neutral SLM methodology put forth by Decisys, Inc. Decisys is a U.S. network-consulting firm that provides analysis and recommendations for corporate managers and technology vendors, including network assessment, strategic planning and migration, and network architecture development.

The SLM methodology recommended by Decisys is described next. Generally, Decisys's SLM methodology closely aligns with ours and corroborates many of our points. There are, however, some interesting and insightful points that we have not mentioned.

The Decisys SLM methodology focuses on the issue of how to set up an SLA. It consists of four major steps:

1. Understand business objectives and user requirements;

2. Establish the baseline of current network performance;

3. Reconcile internal/external metrics and negotiate cost-benefit trade-offs;

4. Write the SLA.

On the surface, it appears that our SLM methodology in Section 3.1 goes into a bit more detail and is more encompassing than the steps outlined here. For example, Decisys's move from step 1 to step 2 requires seven intervening steps in our methodology. However, this is a surface disparity only. If one studies the Decisys literature on SLM methodology in detail, one will see that the steps of our essential methodology are covered adequately.

The particular insights into SLM that Decisys offers are the following:

▶ Implementing an SLA does not have to be an all-encompassing effort that applies to an entire enterprise. It is entirely acceptable and offers significant benefits to start with a single department or a single application. One should decide what service will be defined by the SLA and then baseline that service. Often, separate SLAs are required per autonomous business or functional unit.

▶ One of the biggest challenges in developing SLAs is picking performance metrics that both are measurable and define performance goals that are meaningful (i.e., internal and external metrics). Error rates on lines are measurable internal metrics, but they are not meaningful to application users. The number of application timeouts over a set period of time, which may cause a high error rate, is a more meaningful external metric for users.

▶ If the business is seasonal (e.g., the retailing business), then baselining in February is not going to show much network traffic. If it is not possible to wait for a full year of data collection, then make reasonable estimates and continue collecting data to make ongoing adjustments as necessary.

‣ To get a handle on end-to-end performance, a combination of tools may be necessary, including additional investments in monitoring technology if the existing tools are insufficient to monitor and maintain SLA-based parameters.

‣ Setting up an initial SLA is only the beginning of an ongoing process of improvement and adjustment. Periodic review of the procedures for measuring SLA metrics and review of the metrics themselves will ensure that SLAs continue to meet the needs of the enterprise.

Figure 3.12 shows the SLA review cycle that Decisys recommends. The advice is to set up service level objectives, monitor performance, identify trouble spots, make changes to alleviate or clear up trouble spots (e.g., tune, reconfigure, or upgrade enterprise components), reevaluate service level objectives, and start over.

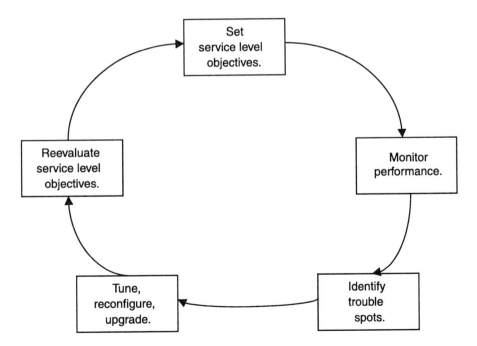

Figure 3.12 SLA review cycle (Decisys).

Note that the Decisys SLM methodology helps to tune an enterprise to obtain maximal performance. That idea adds another useful dimension to our essential SLM methodology.

Summary

This chapter began with a specific, stepwise SLM methodology, which we called the essential SLM methodology. However, it is advised that methodologies in general be adapted and massaged to fit the special characteristics of enterprises and unique talents of practitioners.

Next we argued that SLM system development is a special case of general software development. The chapter looked at proven software development methodologies in the SE community and demonstrated how they apply to our SLM domain. It also looked at use case methodology and the CRC methodology in some detail. The goals of the two methods are to mitigate well-known risks in software projects.

The main thing we found out in our study of software development is that it is rarely a stepwise procedure. We reexamined the essential SLM methodology and suggested several ways to adapt it.

Finally, the case study section discussed an SLM methodology recommended by Decisys. We saw that Decisys's methodology is in close alignment with ours, but in a little less detail. We highlighted some important insights in Decisys's experiences with their clients in SLM development.

Exercises and discussion questions

1. Imagine you are an SLM consultant and you have scheduled a one-hour meeting with a high-level executive of a railroad business. Outline an agenda for the first meeting.

2. Imagine you are a high-level executive of a business. Pick a business with which you are fairly familiar but put all SLM thoughts out of your mind. Pretend you know nothing about SLM. Now, an SLM consultant has managed to arrange a one-hour meeting with you, but you do not expect much to come out

of it. What could the consultant say to stimulate you and pique your interests?

3. In light of the discussion in this chapter, revisit question 4 in Chapter 1.

4. Criticize the sample SLM use case model in Figure 3.3. Are two actors enough to represent users of SLM systems? What other actors (if any) are required? Do our five use cases adequately represent the ways in which the SLM system will be used? What other use cases (if any) are required?

5. This is a research exercise. Figure 3.5 presented event correlation as a black box. Search the literature on event correlation in enterprise management and write a summary five-page paper. *Note:* This exercise will go a long way toward preparing you for Chapter 5.

6. The analysis model in Figure 3.7 shows a database interface object with five arrows going into it. Discuss performance issues (if any) that might arise as we transform Figure 3.7 into a design model. You may want to consult a database expert.

7. Figure 3.10 shows a CRC class card for a trouble-ticketing system. Construct a class card for a system agent, an event correlator agent, and an SLR.

8. This is a research exercise. Search the literature and find a recommended SLM methodology. How does it compare with ours? What additional insights are in it that this book has not covered?

Further studies

In the professional literature, we can find an abundance of material on general SE, but considerably less material on SLM engineering.

The general definition of SE in Section 3.2 is taken from Macro and Buxton's book *The Craft of Software Engineering*. Other useful (and almost classic) SE books include Pressman's *Software Engineering: A Practitioner's*

Approach, DeMarco's *Controlling Software Projects,* Agresti's *New Paradigms for Software Development,* Peter's *Software Design: Methods and Techniques,* and Berzin's *Software Engineering With Abstractions.*

Our discussion of risks in software development is based on the paper by Keil et al., "A Framework for Identifying Software Project Risks" and Glass's paper, "Short-Term and Long-Term Remedies for Runaway Projects."

Our study of the use case methodology is based on the book by Jacobson et al., *Object-Oriented Software Engineering: A Use Case Driven Approach,* and our study of the CRC methodology is based on Wirfs-Brock, Wilkerson, and Wiener's *Designing Object-Oriented Software.* Jacobson and Wirfs-Brock are prominent names in the OO community, but there are others: Booch, Rumbauh, Coad, Yourdan, Shlaer, Mellor, Martin, Odell, Coleman, de Champeaux, Henderson-Selers, and Edwards. A very good comparative study of OO analysis and design methods comes from the University of Twente (www.is.cs.utwente.nl:8080/dmrg/OODOC/oodoc/oo.html). The reader is advised, however, not to get overwhelmed by comparisons of notation and semantics.

As an aside, Jacobson, Booch, and Rumbauh have combined the best features of their respective methodologies into the Unified Modeling Language (UML). Three books on UML (jointly authored by Jacobson, Booch, and Rumbauh) are in press as of this writing.

A very good analytical paper that describes the "starting at the top and the bottom at the same time" strategy outlined in Section 3.3 is the paper by Hegering et al., "A Corporate Operation Framework for Network Service Management."

The case study on Decisys's SLM methodology is based on white papers available at www.decisys.com.

Select bibliography

Agresti, W. (ed). *New Paradigms for Software Development.* New York: IEEE Press, 1986.

Berzin, V. *Software Engineering With Abstractions.* Englewood Cliffs, NJ: Prentice Hall, 1992.

DeMarco, T. *Controlling Software Project.* Yourdon Press, 1982.

Glass, R. "Short-Term and Long-Term Remedies for Runaway Projects." *Communications of the ACM*, July 1998, Vol. 41, No. 7.

Hegering, H-G., S. Abeck, and R. Wies. "A Corporate Operation Framework for Network Service Management." *IEEE Communications Mag.* IEEE Press. Jan. 1996.

Jacobson, I., M. Christerson, P. Jonsson, and G. Overgaard. *Object-Oriented Software Engineering: A Use Case Driven Approach.* Reading, MA: Addison-Wesley, 1995.

Keil, M., P. Cule, K. Lyytinen, and R. Schmidt. "A Framework for Identifying Software Project Risks." *Communications of the ACM,* Nov. 1998, Vol. 41, No. 11.

Macro and Buxton. *The Craft of Software Engineering.* Reading, MA: Addison-Wesley, 1987.

Peter, L. *Software Design: Methods and Techniques.* Yourdon Press, 1981.

Pressman, R. *Software Engineering: A Practitioner's Approach.* New York: McGraw-Hill, 1982.

Wirfs-Brock, R., B. Wilkerson, and L. Wiener. *Designing Object-Oriented Software.* Englewood Cliffs, NJ: Prentice Hall, 1990.

www.decisys.com

www.summitonline.com

www.cabletron.com

www.ics.de

www.comsoc.org

www.acl.land.gov/cgi-bin/doclist.pl

In which we continue thinking about an SLM system as a collaboration among capable agents, much like collaborations we see among capable human beings.

In this chapter:

▶ What is architecture?

▶ The basic SLM architecture

▶ Useful ideas from artificial intelligence, robotics, and data warehousing

▶ SLM architecture revisited

▶ Evaluating SLM proposals with respect to architecture

▶ Case study in telecommunications: Deutsche Telekom

SLM architecture

Chapter 1 served as an introduction to SLM, Chapter 2 laid out SLM concepts and definitions in good systematic fashion, and Chapter 3 described SLM methodology. This chapter discusses the issue of SLM architecture.

First, we answer this simple question: What is architecture? The word *architecture* means different things in different contexts, so we must make clear its meaning as used in this book.

Next, we look at the basic SLM architecture. We show how the basic architecture is flexible enough to be adaptable to enterprises that are network-centric, traffic-centric, system-centric, or application-centric. This chapter takes the viewpoint that an SLM system is best considered a collaboration among intelligent agents. That viewpoint calls for some serious thinking about the concepts of collaboration, intelligence, and data distribution. For that reason, this chapter looks at how those concepts have been developed in the related areas of AI, robotics, and data warehousing.

Guided by our study of achievements in AI, robotics, and data warehousing, we revisit our basic SLM architecture and flesh it out in more detail. Our enhanced architecture provides facilities for distributed management, real-time and off-line SLM, and related management tasks such as trouble ticketing, capacity planning, security control, and software delivery over the enterprise.

We will have a comprehensive SLM architecture in place, but not all SLM systems have to be as comprehensive. Therefore, we will look at ways to adapt and massage the architecture to fit particular circumstances, much like a child's plug-and-play games. We also show how it provides a framework by which to compare and evaluate alternative SLM proposals.

Finally, there is a case study, a very large management system at Deutsche Telekom in Germany. Not all enterprises will be as large as the one at Deutsche Telekom, and not all management systems will be as complex. Nonetheless, the same architectural principles hold for small and medium-sized enterprises, and the Telekom example demonstrates how smaller SLM systems using those principles promise scalability as the enterprise expands.

4.1 What is architecture?

First, let us be clear about what we mean by *architecture*. People often miscommunicate because of the ambiguity of the word. We can easily make it unambiguous in light of the models in software engineering discussed in Chapter 3. Consider the following:

▶ A *conceptual architecture* describes a system at a high level of abstraction without regard to environmental constraints. We show the essential objects in the system and understand how they interact with each other. Compare the discussion of the analysis model in SE in Section 3.2.

▶ A *physical architecture* is the embodiment of some predefined conceptual architecture. Our goal is to carry over the best features of the conceptual architecture into a real working system. In doing that, we strive to retain the best features of the ideal conceptual architecture and have indicators for further research and development. Compare the discussion of the design model in Section 3.2.

▶ A *strategic architecture* can be either a conceptual or a physical architecture, but it additionally shows the phases in which the architecture will be transformed into a working system over time. Compare the discussion of the version N space strategy in SE in Section 3.3.

We are well advised to make clear just what kind of architecture we are talking about in our SLM discussions. For example, given the distinctions in the preceding list, it is easy to see that our discussion of Glaxo-Wellcome in Chapter 1 demonstrates a conceptual/strategic architecture (see Figure 1.5). A good example of a physical architecture was shown in Chapter 3 (see Figure 3.8).

As we describe architectural issues and principles in SLM, we will be explicit about the kind of architecture under discussion. It is important that we do not find ourselves talking about a physical architecture without first understanding a prior conceptual architecture. That would be putting the cart before the horse.

4.2 Basic SLM architecture

Let us consider some fairly well-accepted truths regarding enterprise management:

▶ The enterprise is inherently a distributed, multidomain entity. Enterprises typically are partitioned in ways that help administrators understand and manage them, for example, with respect to geographical domains, functional domains, or managerial domains.

▶ The tasks involved in managing distributed enterprises are too complex for a single agent; thus, the tasks have to be performed by a collection of distributed, cooperative agents.

▶ The data types and data abstractions for monitoring, controlling, and reasoning about enterprise behavior come in various forms, for example, business rules, management data, device data, systems data, applications data, traffic data, user data, symbolic models of the enterprise topology at various levels of abstraction, and events and alarms.

▶ No one vendor can provide all the solutions. Vendors usually are specialists in a few aspects of enterprise management, for example, device, traffic, system, or application management; simulation; help desks; alarm notification systems; and software distribution.

Most people agree that (1) an enterprise management system requires the integration of multiple management products and (2) an enterprise management system should be crafted, built, and developed from an industrywide focus, to which vendors can offer agents that represent their specific areas of expertise.

A useful (and fun) analogy to system integration is the kick-pass play in soccer. One management system (a player) carries a task (the ball) as far as possible, and when the task transforms into a task for which the system is unsuitable, the relevant information is kicked to another system that can carry the task further. Taken to the extreme, system integration can be likened to the dynamics of a soccer game. Other analogies are equally useful. The reader is invited to think about a restaurant, where the maitre d', the waiters, the cooks, and the busboys integrate their tasks toward a common goal.

To corroborate those conclusions, refer to Table 1.2, which shows how the industry has evolved with respect to enterprise management. It is clear from the discussion that SLM builds on existing methods in network, traffic, system, and application management. That fact is evident in the TMN and TINA discussions in Chapter 2. Thus, we would be misguided to talk about SLM (or BP management) without first considering enterprise management. The basic conceptual architecture for SLM in Figure 4.1 illustrates these points.

At this juncture, let us make note of an important disparity in the industry. The management styles of some businesses are network-centric, the management styles of other businesses are systems-centric, others are

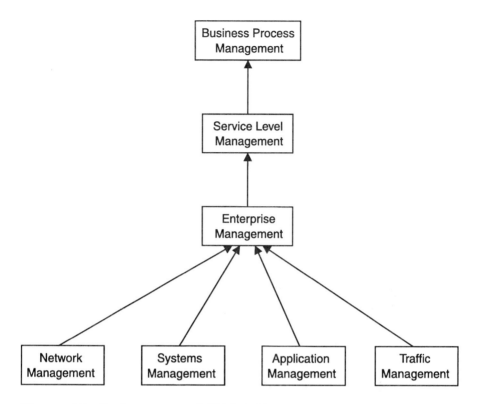

Figure 4.1 Basic conceptual SLM architecture.

application-centric, and still others are traffic-centric. The particular style adopted by a business typically results from two kinds of influences:

- ▶ The business hired a management expert who was well versed in network, systems, application, or traffic management and made recommendations accordingly.

- ▶ The business was convinced by a vendor that one particular area of management was the best way to go.

The type of management is often determined by specific business factors as well. What typically happens is that a business comes to understand that all management areas are important and interdependent. Yet at the same time the business is fairly entrenched in one particular style, using some particular commercial product or suite of products. What does the business do?

It is hard to change styles in midstream. Therefore, the best thing to do is to remain centric with respect to the familiar style unless there are overwhelming reasons to do otherwise. That approach, then, calls for finding ways to integrate other management systems into the existing management system.

The basic conceptual SLM architecture in Figure 4.1 allows for that. For example, suppose the management style of a business is network-centric, using IBM NetView. The business is comfortable with the NetView GUI and how it represents the network topology. Now the business wants to incorporate systems management using Tivoli TME. Thus, NetView escalates to the central enterprise management system, and TME system monitoring agents feed information to NetView, as shown in Figure 4.2. It is the responsibility of NetView to correlate and make sense of both NetView network data and TME system data and to make the data accessible to the SLM agent.

For a different example, suppose a business is application-centric and currently uses Candle products for application management but wants to complement that with Spectrum for network management and Win-

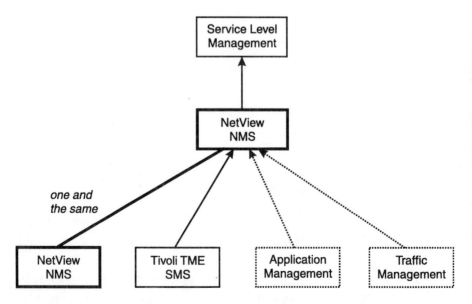

Figure 4.2 A physical network-centric SLM architecture.

Watch for system management. Thus, Candle escalates to the central enterprise management system. Spectrum monitoring agents feed network-related information to Candle, and WinWatch monitoring agents feed systems-related data to Candle, as shown in Figure 4.3. It is the responsibility of Candle to correlate and make sense of Candle application data, Spectrum network data, and WinWatch system data and to make the data accessible to the SLM agent.

Both examples demonstrate the mapping of a conceptual architecture to a physical architecture. It is a good way to think about the possible ways in which multiple agents work together to offer a comprehensive enterprise management system. However, if we dig deeper, we will see that there is considerably more to the story. In Section 4.3, we will think about the following questions:

▶ What do we mean when we say that an enterprise management system is a collaboration among intelligent agents?

▶ What do we mean when we say that an agent is intelligent?

▶ What kind of data exists in the enterprise management domain, and how is it distributed?

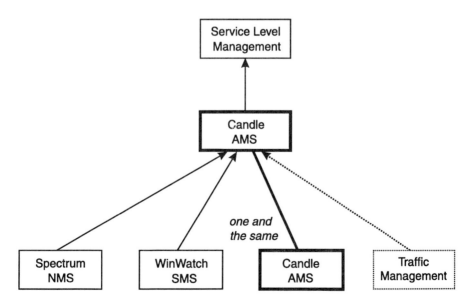

Figure 4.3 A physical application-centric SLM architecture.

Those sorts of questions have been studied in the AI, robotics, and data warehousing communities, although their findings have not been applied to our enterprise management domain in a unified way. Thus, in the following section we will look at the approaches and lessons learned in these disciplines with an eye toward applying their insights to our enterprise management architecture. With these insights, we come back around to flesh out our basic SLM architecture in more detail.

4.3 Useful ideas from artificial intelligence, robotics, and data warehousing

The idea of an intelligent enterprise management system has become popular. Enterprise administrators would like nothing better than to have a relatively autonomous enterprise management system that can perform routine tasks and handle administrative problems reliably with little human intervention. Included would be fault identification and repair; easy configuration of devices, systems, and applications to support the business; identification and correction of performance problems; methods to control the accessibility of enterprise components; and methods to distribute software over the enterprise.

Intelligence in an enterprise management system often is seen as carrying out a business's policies and business rules, with little to no human intervention. To do that, an enterprise management system would have to learn about its current environment and, based on its policies and rules, be able to discern whether an environmental change is problematic or intentional. Learning and proper execution of knowledge are the hallmarks of an intelligence, although they are hard to achieve.

Most seasoned enterprise administrators probably would say this: Keep dreaming and good luck. Nonetheless, that is our goal, albeit an ambitious one. The current thinking is that looking at enterprise management as a collaboration among intelligent agents is the best path to the goal.

For an analogy, consider a group of robots that collaborate on an automobile assembly line. Each robot consists of five items:

▶ A set of sensors by which to perceive the robot's surroundings;

▶ A set of effectors that move the robot around, pick up things, attach one thing to another, and so on;

▶ A communication mechanism by which a robot can confer with other robots;

▶ A reasoning component by which a robot can make decisions based on what it perceives and what it is told by other robots;

▶ A set of policies and business rules that define the goals of their collaborations.

Similarly, an enterprise management system consists of a set of agents, and each agent consists of five similar items:

▶ *Sensors:* device-monitoring agents perceive operating characteristics of devices, and traffic monitoring agents perceive characteristics of network traffic.

▶ *Effectors:* instructions to restrict classes of traffic that flow over network lines, instructions to restrict user access to Web server operating systems, and instructions to download a software application to multiple systems in one fell swoop.

▶ *Communication:* device, systems, and application agents send events to an enterprise agent, an enterprise agent sends an alarm to a paging system, and a paging system sends a message to a trouble-shooter.

▶ *Reasoning:* an enterprise agent studies device, systems, and application events and infers enterprise alarms, enterprise state, and potential bottlenecks.

▶ *Policies and business rules:* all agents keep an eye on the goals of the business when making decisions about which problems to be most concerned about.

In the same way that a group of robots is given a set of policies and the robots collaborate to uphold the policies, enterprise administrators would hope to give the enterprise management system a set of policies whereupon the enterprise management agents collaborate to uphold those policies. For example, an SLA is a kind of policy, and we would expect the agents in an enterprise management system to collaborate to uphold the SLA. We are beginning to see enterprise management in that kind of perspective; as we enter the twenty-first century, we should see a complete shift toward a view of enterprise management as a collaboration among intelligent agents.

That perspective of enterprise management opens the way for thinking about outstanding issues with cooperative multi-agent architectures in general and their contributions to understanding and achieving good enterprise management solutions. Multi-agent architectures have received considerable attention in other disciplines, and similar issues are beginning to surface in enterprise management research and practice.

For example, an important challenge is simply understanding enterprise management tasks with respect to sensing, effecting, communicating, and reasoning requirements over a group of agents. What tasks are better performed by monitoring and reasoning about traffic data? Network and system events? Enterprise alarms? What tasks require a symbolic model of an enterprise and at what level of abstraction? What tasks require a collaboration among a set of agents, as opposed to tasks that require a single agent?

Those are hard questions, and analogous questions have been studied in AI, robotics, and data warehousing. The following sections look at the architectural principles and guidelines that come from those disciplines. At the same time, we consider the problems and goals of enterprise management and explain how the principles in AI, robotics, and data warehousing map to our SLM architecture.

Distributed artificial intelligence

A common approach to problem solving is to decompose a task into subtasks and to perform the subtasks sequentially or assign them to individual agents in parallel. That approach assumes that a central controller exists that decomposes the task, assigns subtasks to appropriate agents, and synthesizes the results.

In contrast, distributed artificial intelligence (DAI), also known as distributed problem solving and multi-agent reasoning, is concerned with problem solving for which a central controller is not present and for which subtasks are interdependent. Specifically, DAI is concerned with problems that exhibit the following characteristics:

▶ Control and expertise required for solving a problem are distributed over a number of agents.

▶ An agent is not always able to complete its subtask alone.

▶ No single agent can solve the entire problem alone.

▶ The solution to the problem requires cooperation and communication among agents in such a way that the expertise of each agent combines to solve the problem.

It is clear that enterprise management is a DAI problem. A first-order categorization of enterprise management tasks includes fault, configuration, accounting, security, and performance management. However, it is not clear that this is the best or a complete categorization of enterprise management tasks, and it is even less clear how and in what ways the agents performing the tasks depend on the other agents.

For example, problem solving that involves fault, configuration, and performance management often requires specialized agents in each area who compare notes to solve the problem. In addition, another agent in budget control and expenditures is consulted to find the most reasonable solution from a cost perspective.

As an aside, it is interesting to note that a problem that could turn out to be a DAI problem is the re-creation of intelligent behavior exhibited by a human agent. For example, a controversial issue in the robotics community is that of determining what mechanisms are required to build an autonomous agent.

One view is that an agent requires a single controller who coordinates tasks such as perception, cognition, acting, planning, and learning. In that view, intelligent behavior is not a DAI problem because a controller delegates those tasks to subagents, and the subagents do their work independently of each other and the controller synthesizes their results.

An alternative view is that intelligent behavior of a single agent is itself an example of DAI. Perception, cognition, and so on, are performed by specialists who negotiate among themselves, and agency is an emergent property or epiphenomenon of the multiple agents.

As with any sufficiently large multi-agent enterprise, including enterprise management and possibly a human agent, those questions are hard. The primary idea we can borrow from the DAI community and apply to our enterprise management problem is how to start thinking about possible architectures for a system of collaborating agents.

Table 4.1 shows eight dimensions of multi-agent systems. The first four dimensions describe possible attributes of the system as a whole,

Table 4.1
Dimensions of Multi-Agent Systems

1. System model	Hierarchical	Heterarchical
	Centralized	Decentralized
	Pure	Mixed
2. System grain	Fine	Coarse
3. System scale	Small	Large
4. System adaptability		
Before problem solving	Low	High
During problem solving	Low	High
Re: environmental change	Low	High
5. Agent autonomy	Controlled	Autonomous
6. Agent resources		
Bandwidth	Restricted	Ample
Memory	Restricted	Ample
Cohorts	Many	Few
Local knowledge	Smart	Dumb
Global knowledge	Smart	Dumb
	Detailed	Abstract
7. Agent communication		
Form	Message passing	Shared memory
	Direct	Indirect
	Selective	Broadcast
	Unsolicited	On-demand
	Push	Pull
	Acknowledged	Unacknowledged
	Single-transmission	Repeated
	Protocol	Free-form
Content	Partial	Complete
	Certain	Hypothetical
	Detailed	Abstract
	Highly structured	Unstructured
	Benevolent	Malevolent
8. Agent type	Uniform	Heterogeneous
	Human	Software

while the last four dimensions describe possible attributes of agents that compose the system.

Dimensions of multi-agent systems

System model   A system model is the organizational structure of the agents in a DAI problem. At one extreme, the structure can be hierarchical. Agents at each level of the hierarchy can communicate with agents on the levels above and below it but not with agents on the same level. At the other extreme, the structure can be heterarchical. Agents are considered to be peers on the same level. In a mixed hierarchical-heterarchical structure, some agents are allowed to communicate with peer agents and also subordinate and superordinate agents.

In a centralized model, one agent (or a cluster of agents) makes all control decisions, although the decision can derive from negotiations with other agents. In a decentralized model, there is not a central decision maker that has the final word. Decisions are a collaborative effort.

Models can be mixed or relatively pure. For example, a hierarchical model with one centralized decision maker at the top of the hierarchy is a pure model. Obviously, a decentralized hierarchical-heterarchical model is mixed.

System grain   The system grain is the comparative size of the agents that make up the system. In our enterprise management domain, an example of a fine-grained system is one that comprises a large number of device monitoring agents, where the number of monitoring agents is equal to the number of devices. An example of a coarse-grained system is one in which there are a small number of enterprise monitoring agents such that a single enterprise agent monitors a large number of devices.

System scale   System scale refers to the number of agents in a system and thus is related to system grain. An enterprise management system that comprises hundreds of separate device monitoring agents is a large-scale system, whereas an enterprise management system that manages the same domain with a dozen enterprise agents is a small-scale system.

System adaptability   System adaptability refers to the degree to which a system can modify itself. One sense of adaptation is the ability of agents in the system to restructure themselves when assigned a particular task. A second sense of adaptation is the ability of a system to continue to function when part of the system fails (known as graceful adaptation). The ability and degree to which a system can adapt to environmental changes largely depends on its model, grain, and scaling factors.

Agent autonomy   Agent autonomy refers to the ability of an agent to make its own decisions. An agent that simply carries out assigned tasks is controlled. An agent that is offered tasks and decides whether to take them is semiautonomous, that is, mixed controlled-autonomous. An agent that reasons about, formulates, and assigns tasks is autonomous.

Agent resources   Agent resources is a large category that refers to the physical and "mental" characteristics of an agent. With respect to physical character, resources include communication bandwidth, CPU and memory constraints, and access to other agents (i.e., cohorts) in the system. With respect to mental character, resources include an agent's special expertise, an agent's knowledge about the expertise of other agents, and an agent's knowledge about the organizational structure of which it is a part.

Agent communication   Agent communication is also a large category. We distinguish between the form of a communication and the content of a communication. With respect to form, two agents might communicate via message passing or via a common shared memory (the latter is sometimes called a blackboard system). With message passing, two agents might communicate directly or indirectly through an intermediary agent. An agent can transmit a message to a select group of other agents or broadcast a message to the whole organization. A message received by an agent may be unsolicited, or it may be have been demanded beforehand by the agent. That sometimes is referred to as push and pull technology, respectively. An agent can acknowledge or not acknowledge receipt of a message. A message can be transmitted just once or retransmitted at intervals (also known as murmuring). Finally, communication among a collection of agents may proceed according to a well-defined protocol, or it may proceed in an ungoverned, free-form style.

    With respect to the content of a communication, a message can be partial or complete, certain or hypothetical. The message can be detailed or abstract. The syntax of the message can be well structured or in a free-form style. If it is free form, the recipient agent is burdened with parsing the message and figuring out what it means, which implies strong reasoning abilities. Finally, the intent of the message typically is assumed to be benevolent; however, some researchers argue that this is an unfounded assumption about organizational behavior that hides important issues in group dynamics.

Agent type   Agent type refers to the kinds of agents in a system. A system's agent type is uniform if all the agents in the system are of the same kind. For example, if an enterprise management system consists only of monitoring agents, then the agent type is uniform. Otherwise, the system consists of heterogeneous agents (known as hybrid systems). If the enterprise management system also includes an event correlation agent and an alarm notification agent, the system is heterogeneous. Finally, the agents can be human agents or software agents.

General rules

Our understanding of the possible system attributes and agent attributes in Table 4.1 provides us with a good framework for conceptualizing, designing, and developing enterprise management systems. Some general rules that we can apply to the analysis and design of enterprise management systems are the following:

- ▶ Distribute intelligence over agents as evenly as possible, that is, do not have a single agent doing all the work or carrying most of the load.

- ▶ Assign intelligence to match the level of an agent's task. In other words, do not give the agent more or less intelligence than required for performance of its task.

- ▶ Agents that perform huge amounts of internal processing should not be required to communicate with other agents extensively.

- ▶ Agents that communicate with other agents extensively should not be allowed to perform huge amounts of internal processing.

- ▶ To increase system adaptability, try to avoid wholly centralized architectures.

▶ The fewer agents one can use to get the task done, the better. (Compare the discussion of Ockham's Razor in Section 2.2.)

It is interesting to note how the guidelines from the DAI community complement the guidelines we learned from the SE community in Chapter 3. For example, the distinction among interface objects, entity objects, and control objects and our warning to keep those objects distinct are echoes of the guidelines to distribute intelligence evenly and to keep an agent's local knowledge specialized.

Reflect for a moment about human collaboration, including personal experiences and observations of remote human collaborations. Why is it that some collaborative efforts turn into fiascoes, while other collaborative efforts seem to come together smoothly to achieve a common goal? What properties distinguish the former from the latter?

Sociologists, political scientists, philosophers, and other thinkers have studied those kinds of questions for 2,000 years. It is interesting that we now have to consider similar questions in a highly technical context. That is an example of anthropomorphism: the imposition of human and social qualities on nonhuman entities. As a rule of thumb, anthropomorphism is a good way to think about complex systems, especially during the construction of a conceptual architecture. It is a common, recommended principle in current software engineering practices. However, the physical embodiment of the architecture may begin to lose some of its human qualities. The classic example is an airplane. Airplanes certainly fly but not like birds and butterflies.

Robot intelligence

The work begun in the 1970s on implementing intelligence in robots is an interesting, ongoing story. If our hypothesis that enterprise management will come to be viewed as a collaboration among intelligent agents turns out to be true, then the uncovering of false starts and promising approaches in the robotics community could help us in understanding intelligent enterprise management.

This section describes three approaches to the development of intelligent architectures for robots:

▶ Single-loop architecture;

▶ Multiloop architecture;

▶ Subsumption architecture.

The *single-loop architecture* for robot intelligence is shown in Figure 4.4. Intelligent behavior starts with robot sensors and ends with instructions that are executed by robot effectors. The flow of information begins with the abstraction of sensory input (going up the left side of the figure), reasoning (going from left to right at the top), and instruction (going down the right side).

Initially, sensory input is passed through several layers of abstraction. Typically, the levels of abstraction are signals, signs, and symbols. The initial set of signals over time can be portrayed as the graph shown at the bottom left in Figure 4.4. Going up the left side of the illustration, we see that the signals are transformed into signs, such as a sequence of 1s and 0s by an abstraction module A1. Finally, the sequence of signs is transformed into the symbol "obstacle 20 degrees right" by A2. Note that each abstraction module filters noise and extraneous data out of the data passed to it and transforms the data into smaller, but more informative, chunks of knowledge.

When information becomes manageable at the symbolic level, it is compared with predefined knowledge about what instructions should ensue. That operation usually is defined as "reasoning" and can be

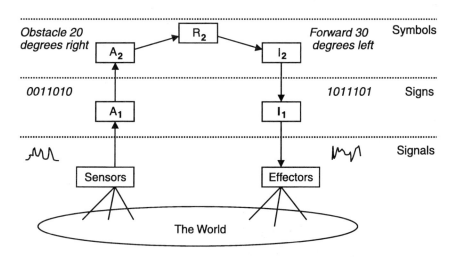

Figure 4.4 Single-loop architecture for robot intelligence.

implemented in a number of ways, including simple lookup tables and rule-based expert systems. For example, consider this rule:

> IF an obstacle is X degrees to the {left, right} and X is less than 25
> THEN move forward ($X + 10$) degrees to the {right, left}.

Thus, the output of the reasoning module R2 is the symbolic instruction "forward 30 degrees left." Now, that instruction is decomposed down through the same levels of abstraction by instruction modules I2 and I1 until it is finally suitable for processing by the robot's effectors.

The single-loop architecture is sometimes called the classical architecture for robot reasoning, and it was the first approach to implementing robot intelligence in the late 1970s. However, it did not work. Consider the following.

Sensor data must be abstracted into symbols before it can be coordinated with the goals of the system and initiate the decomposition of instructions for effectors. As a result, the processing time required to transform sensor data to control instructions through the loop is prohibitive. By the time the system figures out what to do in environment A, it is likely that environment A is obsolete and the instructions are no longer applicable. In other words, the rate at which the world changes over time may be disproportionate to the amount of processing required for timely behavior.

Robotics researchers observed that the reasoning does not have to take place only at the symbolic level. Reasoning can occur at all levels of abstraction. Further, they observed that reasoning at lower levels of abstraction require increasingly less time, because upper levels are bypassed. Therefore, problems that are time sensitive and require an immediate response should be handled at lower levels, and problems that are not so urgent should be handled at upper levels.

The distinction generally is referred to as *reflexive behavior* for short-term problem-solving and *deliberative behavior* for longer-term problem solving. Figure 4.5 shows a simple but important enhancement to the single-loop architecture that reflects those ideas, the *multiloop architecture*.

Each level of the multiloop architecture is a separate control loop that corresponds to a specific class of problems, where problems are parti-

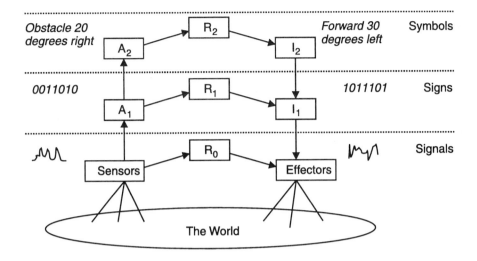

Figure 4.5 The multiloop architecture for robot intelligence.

tioned and assigned to levels according to the amount of time and type of information required to solve them.

For example, the short-term abstraction-reasoning-instruction loop at the lowest level provides quick reaction, bypassing upper level control mechanisms. In our enterprise management domain, such tasks might include temporary disconnections of a busy server or an immediate action to switch to a backup server in the event of failure of a primary server. Another example is traffic shaping to support integrated multimedia services such as voice, data, and video on demand.

The medium-term loop provides reaction to more complex problems and operates on increasingly abstract data. In our domain, tasks of that sort might include event correlation in a busy enterprise with multiple "contact lost" events, when some particular event is the real culprit and other events are effects of the culprit event. The instruction might be to forward an explanation and recommended repair procedures to a repair-person via a pager or to actually initiate the repair procedure automatically.

The top level would provide reaction to problems that are less urgent and allow more time for performing an analysis. The classic example of such a task is the reasoning involved in deciding to move a host from subnet A to subnet B because the majority of the host's clients reside on subnet B, thereby causing increased traffic on the link between A and B. Another example is long-term capacity planning.

In comparison with the single-loop architecture, the multiloop architecture is intuitive and shows promise for a clearer understanding and design of enterprise management systems. However, there remain challenges:

▶ It is not straightforward to classify and divide tasks according to information available and time required to solve them.

▶ It is not straightforward to design the particular algorithms that reside in the abstraction, reasoning, and instruction modules.

▶ It is not clear how the modules on multiple levels cooperate when a task requires multiple levels of reasoning. This is the classic DAI challenge.

The *subsumption architecture* was conceived in reaction to the problems of the earlier architectures and is studied and developed for various kinds of intelligent robot systems, including mobile, underwater, and walking systems. If we look at the single-loop architecture and multiloop architecture in Figures 4.4 and 4.5 we see that the system is decomposed into functional modules. Sensor information flows through the modules of the architecture until it is transformed into a raw control instruction.

In contrast, the subsumption architecture avoids such a modular decomposition. Instead, the subsumption architecture engineering approach is to decompose a task into a collection of simpler task-achieving behaviors that are tightly bound together (Figure 4.6).

The behaviors reside on levels such that:

▶ Higher levels exhibit increasingly complex behaviors.

▶ Each level subsumes (i.e., uses) the behaviors on the levels beneath it.

▶ Lower levels continue to achieve their level of performance even if the level above fails.

Unlike the single-loop and multiloop architectures, sensor data is not transformed through levels of abstraction. Instead, all levels monitor all sensor signals, and certain combinations of signals trigger appropriate behaviors. The output of a level-N behavior interferes with the output of

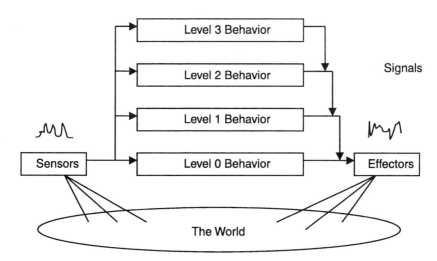

Figure 4.6 The subsumption architecture for robot intelligence.

levels beneath *N* to produce an enhanced behavior. In that way, some kind of behavior is possible even when upper level-*N* behavior is disabled.

The subsumption architecture has been demonstrated on robots with a few simple behaviors. An example is path traversal in a space that contains unpredictable moving objects. Suppose the goal of the robot is to get from point A to point B without suffering damage from collisions with foreign objects. A level-0 behavior is to remain still but to dart away in a random direction on the sensation of a foreign object. A level-1 behavior is to dart away but in the general direction of point B. Finally, a level-2 behavior is to move toward B if the path is clear of foreign objects. Now, to confuse the robot, let's assume that we take away the robot's knowledge of B. Then, although the level-2 and level-1 behaviors are dysfunctional, the robot is supported by its level-0 behavior until it gets its bearings straight.

Consider this example in our enterprise management domain: Suppose a server monitoring agent sees all server events and is capable of identifying bad events and forwarding them to a repair person via pages. Further suppose that we have a very large number of such agents monitoring a Web server farm. That is a level-0 behavior, and it is not hard to build agents to do it.

Now consider an enterprise agent that sees all server events and all device and systems events. The job of the enterprise agent is to do event correlation over three varieties of events. That is a level-1 behavior. The agent should be able to figure out the root cause of a collection of bad events having to do with servers, network devices, and systems.

For example, if the agent reasons that a multitude of bad server events is really an effect of a failed networking device, then the agent interferes with the level-0 behavior. The output of the level-1 behavior is to suppress the forwarding of numerous server and application events and instead to forward a single device event to the repairperson.

The beauty of the subsumption architecture approach is that even though the level-1 behavior might become dysfunctional, we still have some kind of management happening. If the level-1 behavior were to fail, then the repairperson would be flooded with pages regarding server and application malfunctions. However, that is better than nothing. The burden of event correlation is shifted from the enterprise management system to the repairperson.

In general, the arguments in favor of the subsumption architecture approach are as follows: Once a lower level control system is in place and operable, it is not altered again. Higher levels may enhance the output of lower levels, but they do not interfere with the levels' internal structures. Thus, the subsumption architecture allows experimentation with new behaviors without interfering with mechanisms already in place.

In addition, there is not a symbolic layer in the architecture. The hallmark of the subsumption architecture is that the world represents itself, rather than a symbolic model representing the world. The world is represented via continuous unabstracted sensor input, and behavior is interleaved intimately with goings-on in the world. Thus, there is not a lag in response to the world while the system is reasoning about what to do.

The arguments against the subsumption architecture approach are these. Scaling the subsumption architecture can be prohibitive. The addition of increasingly complex behaviors can result in unpredictable behavior. Also, a methodology for analyzing a complex behavior into simpler behaviors and combining them in the right ways is not available, nor is there a methodology for reconfiguring a collection of existing simple behaviors to achieve other complex behaviors.

In sum, the verdict is still out regarding a correct architecture for robot intelligence. The two main camps are the symbolic camp (favoring the multiloop architecture) and the subsymbolic camp (favoring the subsumption architecture). The former is usually identified with Carnegie Mellon University (CMU) and the latter with the Massachusetts Institute of Technology (MIT). Other researchers are trying to construct a unifying architecture that incorporates the best features of both camps. The debate on the topic is interesting and often heated. Nonetheless, we will take advantage of the ideas and concepts in this work when we revisit our SLM architecture in Section 4.4.

Data warehousing

This section switches gears and examines architectures used in data warehousing. The idea of a data warehouse is not new. Methods in data warehousing are well established in many domains besides enterprise management, including financial analysis, marketing, and insurance.

The basic idea is to collect data from several sources, clean it up, put it into a data warehouse, and then perform analyses on the data with specific goals in mind. Imagine a company putting sales data into a data warehouse, where the data includes purchased items and geographical characteristics of the purchaser. A business can perform a clustering analysis that shows the kinds of items bought in particular geographic regions. The business then can penetrate new markets in similar geographic regions with the same kinds of items.

It is straightforward to see how the concept of data warehousing is applicable to enterprise management. We want to collect performance data issuing from several monitoring agents into a data warehouse. With such historical data, we can perform analyses regarding usage trends, configuration modifications to increase performance, strategies for expanding the enterprise, accounting, and service level reporting.

The key concepts in data warehousing are operational data, data scrubbing, the data warehouse, and data marts (Figures 4.7, 4.8, and 4.9).

▶ *Operational data* is data collected at a source, where the source is close to the operation of the enterprise. Examples are monitoring agents such as Spectrum enterprise agents, WinWatch system agents, Patrol application agents, Netscout RMON traffic agents, and special-purpose data-collection agents. Because operational data is close to the source and is at a low level of abstraction, it can be used for real-time tasks such as alarming and time-sensitive control. Figures 4.7, 4.8, and 4.9 illustrate three enterprise agents that

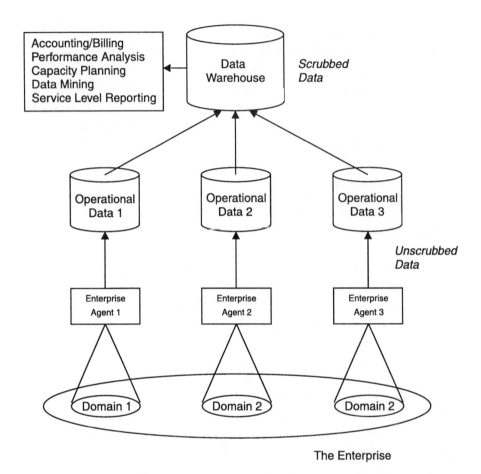

Figure 4.7 A data warehouse scheme with one warehouse.

monitor three geographical domains in a large enterprise, producing unscrubbed data for each domain.

▶ *Data scrubbing* is the process of cleansing operational data in preparation for moving it to a data warehouse. Examples of data scrubbing are (1) replacing a garbage value with NULL, (2) collapsing duplicated data, and (3) filtering out irrelevant data.

▶ A *data warehouse* is the repository where scrubbed data is put. Typically, the data warehouse is implemented in a commercial database system such as Oracle or Microsoft SQL Server. Many data warehouses include reporting facilities and generic algorithmic

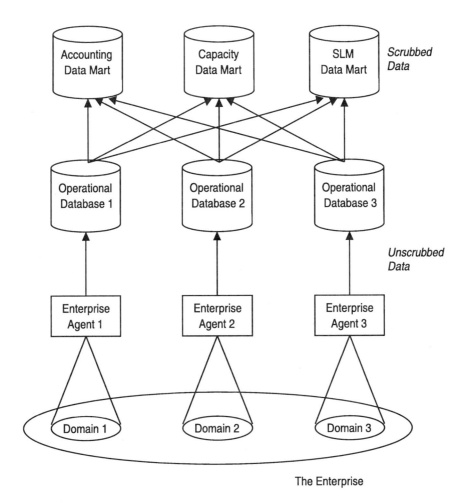

Figure 4.8 A data mart scheme, functionally distributed.

methods for analyzing the data, for example, Crystal reports and data mining algorithms.

▶ A *data mart* is a collection of repositories where scrubbed data is put. Usually, a data mart is smaller than a data warehouse and holds specialized data suited for a particular task. For example, a data mart might exist solely for holding accounting data, another data mart for holding data to perform capacity analyses, and another for holding data for service level reporting.

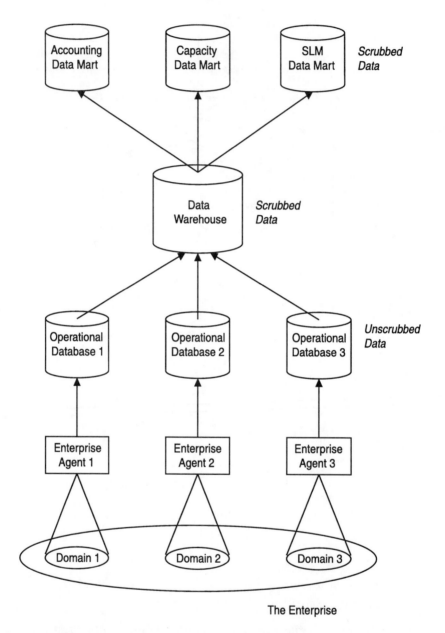

Figure 4.9 A combined data warehouse scheme and a data mart scheme.

In general, there are a number of schemes by which to distribute data so that it is easily accessible by the right application, with minimal communication and performance costs. One option is to configure enterprise monitoring agents to forward select data directly to special-purpose data marts, as shown in Figure 4.8. Another option is first to collect all data in a central warehouse and then distribute it to data marts for special-purpose tasks, as shown in Figure 4.9.

At this juncture, a general theme is coming into focus. Particularly, we can make a distinction between two modes of enterprise management:

▶ Real-time enterprise management close to data collection sources;

▶ Off-line enterprise management relatively far from data collection sources.

Real-time enterprise management happens at low levels of abstraction and is performed by monitoring agents. Such tasks include local event correlation, alarming, and time-sensitive control of the enterprise processes.

Off-line enterprise management happens at higher levels of abstraction and is performed by agents that are less restricted by time-sensitive decision making. Such tasks include accounting and billing, capacity planning, service level reporting, and general data mining with specific goals in mind.

Real-time agents perform in the present, while off-line agents support the future. Real-time agents maintain the environment on a daily basis, whereas the off-line agents serve to mature and direct environmental changes for the future.

Clearly, real-time and off-line enterprise management are interdependent. A simple example will suffice to show that. Suppose the SLM methodology in Chapter 3 has been worked through properly. Services have been identified, services have been mapped to components, the SLA is firm, and the component monitoring agents are in place. The agents are monitoring their respective component parameters and passing values to a data warehouse. At the end of each month, the supplier and consumer plan to check the SLM reports against the service agreement.

Now, the supplier would like to know early on whether it is likely that the terms of the agreement will be met and whether things can be corrected if it appears that the agreement will be violated. Further, the supplier would like to know immediately if a hard fault occurs that will compromise the agreement. Thus, we have two important modes of SLM: real-time SLM and off-line SLM. The former will help ensure the success of the latter.

4.4 SLM architecture revisited

With the ideas and insights of Section 4.3 under our belts, let us depict our SLM architecture in more detail. Figure 4.10 is an enhancement of the basic architecture shown in Figure 4.1. One difference is that the off-line management system (OMS) at the top replaces the SLM system. By that, we simply mean that an SLM system is one kind of OMS, and there may exist other kinds of OMSs.

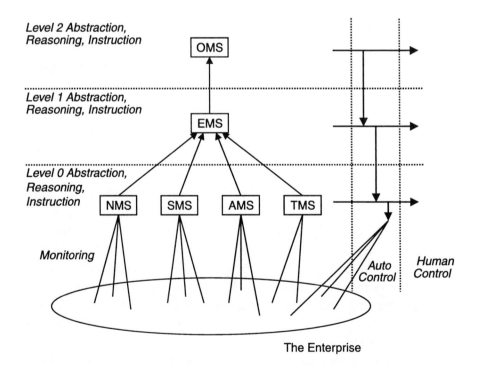

Figure 4.10   Conceptual enterprise management architecture.

Consider fault management. Monitoring systems at level 0 identify faults in their areas of expertise, whereupon they issue control instructions. A control instruction may be to execute an action directly on an enterprise component (unsupervised control), to log the fault in a trouble-ticketing system (supervised control), or to pass the fault to an enterprise management system on level 1.

The enterprise management system on level 1 reasons about faults across individual areas of expertise and may issue similar instructions. The classic example of a level-1 behavior is the performance of event correlation over network, system, application, and traffic events.

The off-line fault management system at level 2 may analyze faults from a historical perspective, with the goal of discovering trends that are hard for systems on level 0 or level 1 to detect. A good example of a level-2 behavior is the execution of a data mining algorithm to try to discover what general enterprise conditions lead to certain classes of faults.

Now let us consider the main topic of this book, SLM, in light of the discussion of fault management. If we have followed the methodology in Chapter 3 properly, we will have identified the components and component parameters that offer a service. Thus, an off-line SLM agent on level 2 should know whether a particular component contributes to the health of a service and take action accordingly whenever the component fails or begins to degrade. Assuming that a good fault management system is in place, it is rather straightforward to extend the system into the SLM space.

The structure and the rationale of our architecture should be agreeable for the most part. However, there is likely to be the lingering question of how we carry it over to a physical architecture. Plainly put, how do we embody it and make it work? Our case study at the end of this chapter shows how some engineers are making it work. But first let us show how we can evaluate SLM proposals with respect to our architecture.

4.5 Evaluating SLM proposals with respect to architecture

The architecture in Figure 4.10 provides a good framework by which to compare and evaluate alternative SLM proposals. This section provides a series of questions and answers regarding an SLM proposal. For businesses that depend on the enterprise, it is a good set of questions to ask vendors and service suppliers. And for vendors and suppliers, it is a good set of questions to have answers to.

Consider these 10 questions as a final exam for a university-level course on SLM, and each question is worth 10 points.

1. Does the system do componentwise monitoring and reporting?

    Answer: The answer should be yes, but that is not sufficient for SLM. This happens when a monitoring agent simply collects data and deposits them directly into a data repository, whereupon the system generates reports over the data. If the answer is merely yes without some indication of other requirements for SLM, take away points.

2. What classes of data—network device, traffic, system, or application—does the system collect?

    Answer: In principle it can collect all classes of data, but not all services will require all classes of data. The definitions of the services and the underlying components on which they depend instruct us as to what classes of data we collect. If the answer is merely that it does indeed collect all classes of data, take away points.

3. Does the system do both real-time and off-line SLM? That is, do we have a way of indicating early on whether an SLA will be met or violated?

    Answer: The answer should be yes.

4. Does the system have an enterprise agent that correlates data received from multiple monitoring agents?

    Answer: The answer should be yes.

5. What method does the enterprise agent use to correlate data received from multiple monitoring agents?

    Answer: There likely will be different kinds of answers to this question, because this is an ongoing area of research. Typical answers are (1) a rule-based expert system, (2) a model-based,

OO reasoning system, (3) a case-based reasoning system, and (4) a belief network. (Chapter 5 describes the strengths and weaknesses of each of these methods.) If some method is named, accompanied by a reasonable explanation of the method, give full credit. If a comparison with alternative methods is explained, give extra credit.

6. Suppose a router hosts a Web server farm and an ATM switch fabric hosts a number of such routers. Now suppose an ATM switch on the edge of the fabric fails, which in turn causes a portion of the farm to be unreachable. How would the system handle that scenario? How many alarms are generated?

   Answer: The answer should be one major alarm on the ATM switch. The alarm should be forwarded to an administrator long before users of the Web farm begin to experience difficulty. However, the way in which the system figures that out should be explained with reference to the answer to question 5. If it is not, take away points.

7. Does the system provide options for notifying an administrator when a hard SLM fault occurs?

   Answer: The answer should be yes (the more options, the better). Answers may include paging, e-mail, trouble tickets, or sirens.

8. Does the system provide mechanisms for automated control of enterprise components and resources?

   Answer: In principle, yes, but one must beware of giving the enterprise management system too much autonomy. We can distinguish between supervised control and unsupervised control. In supervised control, the system asks an administrator if it can execute an instruction on some component, along with an explanation of the symptoms of the problem and why the instruction will repair the problem. This answer can be tied to question 7; if it is, give extra points.

9. How does one add new knowledge to the SLM system, including additional knowledge for the monitoring agents and the enterprise agent?

Answer: First, the answer should be tied closely to the method described in question 5. If it is not, take away several points. For example, if the method is a rule-based system, the answer should illustrate how a new rule adds more knowledge to the system. Second, the addition of new knowledge should be easy, something someone without a master's degree in computer science can do in the field.

10. Does the system have the ability to learn knowledge and adapt to a changing enterprise?

    Answer: This is the most difficult question of all; any reasonable attempt at an answer is acceptable.

Note that the sequence of questions has to do with the conceptual architecture. When we begin thinking about the embodiment of the architecture, a multitude of technical questions will arise. Can the system manage such-and-such kind of device, system, or application? Can the system manage components with such-and-such protocol? How are the agents in the system integrated (e.g., via SNMP, CORBA, NFS, or remote procedure calls)? Will the system scale as the enterprise expands? What resources are needed to implement the system (e.g., UNIX, NT, both, memory requirements)? Is the system Web accessible? What does an SLA look like? What does an SLR look like? How much does it cost for initial deployment and yearly maintenance? Is the performance of the system guaranteed? And so on.

Those questions are necessary, and they reflect the move from our ideal conceptual architecture to a physical embodiment. However, following sound engineering practices, it is important that the conceptual architecture is fairly well understood before the move is made.

4.6 Case study: Deutsche Telekom

This section describes an embodiment of the architecture at Deutsche Telekom (Germany). The enterprise management system is network-centric. The enterprise management agent is Spectrum, built by Cabletron Systems (United States). The system is very large scale, and employs upward of 4,000 enterprise management agents. We show how SLM is built on top of the enterprise management system and discuss ways to extend it further into the SLM space.

First, we will say a few words about the architecture of Spectrum and then describe how Cabletron tested the performance of the system in preparation for moving it to the field.

The Spectrum enterprise management platform is based on a distributed client/server architecture (Figure 4.11). The Spectrum servers, called SpectroSERVERs (SSs), monitor and control individual domains in the enterprise. The Spectrum clients, called SpectroGRAPHs (SGs), may attach to any SS to graphically present the state of that SS's domain, including topological information, event and alarm information, and configuration information.

The SGs are examples of pure interface objects, while the SSs are examples of hybrid interface-control objects (see Chapter 3). The SGs are the interfaces to the enterprise administrators, but they do not have direct access to the enterprise. The SSs provide the interface to the enterprise, but they are not responsible for displaying data; they pass data to the SGs for display.

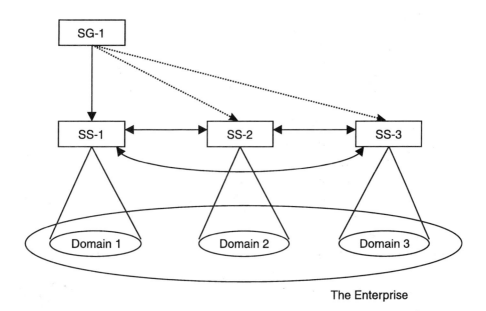

Figure 4.11 Spectrum's distributed client/server architecture.

Any domain can be viewed from a single SG. If SG-1 is in communication with SS-1, but the user wishes to monitor and control the domain covered by SS-2, the user can click on an icon in SG-1 that represents SS-2. Figure 4.11 shows a primary client/server communication between SG-1 and SS-1. Virtual communications between SG-1 and other SSs are indicated by dotted lines.

To prepare for deployment of the system at Deutsche Telekom, Cabletron tested the physical architecture in an internal testbed to establish operational feasibility. A virtual enterprise network was configured using 250 SSs situated in Cabletron facilities in the United States and England. A three-layered hierarchical topology was used, with one master SS connecting to 14 SSs, each of which in turn was connected to 15 to 20 more SSs. Each end-node SS monitored several hundred manageable devices. A total of 15 SGs were attached to each SS at the two top layers of the hierarchy, and each SG was given permission to drill down to the end-node SSs on demand.

Cabletron recommends a maximum 1:7 ratio among SSs that are configured hierarchically. That ratio is derived from workstation operating system characteristics rather than communications traffic load among SGs and SSs. Note that load would be a strong, adverse factor if we allowed a one-to-one correspondence between SGs and SSs, or if we allowed heavy communication among SSs.

In the testbed, Cabletron exceeded the 1:7 ratio to stress the architecture. Performance, accuracy, and reliability were monitored over a period of several weeks, during which time a variety of simulated failures were introduced and the resultant behavior analyzed. A few communication flaws at low-level physical layers were identified and corrected, but as a whole the test was successful and the overall system functioned as planned without incident.

The initial tests provided a good, empirical argument for the scalability of the distributed, client/server architecture. Each SS is an intelligent domain-monitoring agent, capable of presenting management data on demand to any client SG. That keeps inter-SS communications to a minimum. Each SS knows about its peer SSs but is prohibited from extensive communication with them. Later, it is demonstrated how SSs can communicate by intermediary agents that reside at a higher level of abstraction.

The distributed version of Spectrum has been installed at dozens of customer sites worldwide, with setups ranging from a few (two or three) SSs to several hundred. For the most part, customer enterprises are divided into geographical domains, and an individual SS monitors and controls each domain. A central master SS typically is located at a business's headquarters. That sort of arrangement allows follow-the-sun management for global enterprises, where client SGs alternatively attach to a master SS to take over the control of the global enterprise.

The deployment of Spectrum to manage the telecommunications network in eastern Germany is particularly challenging because it relies on a non-SNMP proprietary communication protocol. Spectrum was selected as the management platform for four reasons:

▶ It has a distributed client/server design, thus promising scalability.

▶ It includes tools and procedures for developing non-SNMP management applications.

▶ It includes tools for configuring intelligent SS agents with respect to the flow of management data.

▶ It enables representation of both components and services.

Consider the third item in the list. With multidomain enterprises with corresponding SS agents, polling-based management can be costly in terms of bandwidth load. By restricting SS polling (i.e., using it only for testing basic element presence or status) and instead having managed components forward data to the SSs via traps, in-band management traffic is reduced considerably. That was a requirement for deployment. As an aside, note that a transition from polling to trap-based management via intelligent agents is considered by some people to be the future of enterprise management and is a controversial topic.

Consider the fourth item in the preceding list. Data collected via the enterprise management system is utilized in two ways. First, network devices in all domains are represented topologically to monitor and control the operations of the telecommunications network as a whole. Alarms are generated for devices that experience outages and degradation. Spectrum's event correlation capability prevents the problem of alarm flooding. An example of the alarm flooding problem is when a particular failed device causes apparent, nonreal alarms on a large number of other devices.

Second, the total collection of device alarms is mapped into a well-defined service agreement made with high-profile customers for whom the telecommunications network is crucial. The service agreement states that repair procedures for alarms that affect high-profile customers are given a higher priority than are alarms for lower-profile customers. That is accomplished operationally by assigning relative weights to high-profile and lower profile alarms. At the end of the month, it is an easy matter for both supplier and consumer to view the total collection of alarms and determine whether the agreement has been met or violated.

By the end of 1996, the number of SSs deployed for this enterprise management system was close to 1,000 (490 primary servers, each with a fault-tolerance backup server), with no essential performance problems. As of the publication of the book, 1,700 SSs have been deployed, likewise with no essential performance problems.

The project is on track toward a planned deployment of 4,000 SSs shortly after we enter the twenty-first century. The success achieved thus far has resulted in commitments for similar projects by telecommunications providers in several other countries.

A further consideration: Multiple layers of abstraction

Stop a moment to think about the management tasks that occur within domains and management tasks that occur across domains. We will use fault management as an example.

Fault management at Deutsche Telekom consists of event monitoring, event correlation, event-to-alarm mapping, diagnosis and repair of causes of alarms, alarm-to-service mapping, and service level reporting with respect to the repair of high-profile and low-profile alarms.

Each SS performs those tasks with Spectrum's event correlation mechanism and alarm reporting facilities (Chapter 5 describes Spectrum's event correlation technique and alternative techniques). We can refer to that as *intradomain* event correlation and alarm reporting.

With large multidomain enterprises, the requirement now is to perform the same function across domains. For example, an alarm on a failed router in domain 1 may affect the applications running in domain

2. Conversely, the cause of an application failure in domain 2 may be identified as the result of an alarm on a failed router in domain 1. We refer to that as *interdomain* alarm correlation and alarm reporting.

Thus, we have processes going on at three levels of abstraction: (1) event correlation and alarm reporting with respect to individual domains; (2) alarm-to-service mapping and service reporting with respect to individual domains; and (3) alarm correlation across multiple domains. In simple terms, individual SSs have local knowledge and reasoning capabilities with respect to their domains of interest, but they do not have global knowledge of the entire enterprise—they cannot see the forest for the trees.

Because the physical architecture permits only limited intercommunication among SSs, we need to find some other way to perform the interdomain alarm correlation task. In light of the conceptual architecture in Figure 4.10, we can understand the interdomain alarm correlation task as shown in Figure 4.12.

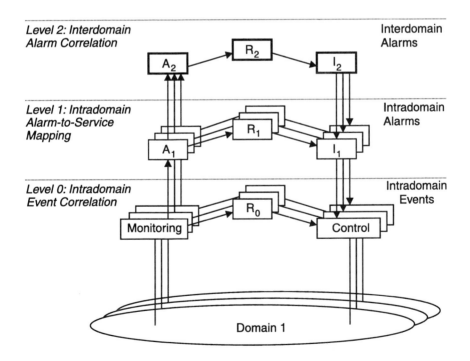

Figure 4.12 A multilevel architecture for multidomain fault management.

The bottommost levels 0 and 1 are performed by SSs that monitor and control individual domains in the enterprise. The agent that resides on the top on level 2 collects alarms from multiple SSs and carries out interdomain alarm correlation, communicating with other SS agents on lower levels as appropriate. Note, then, that the SS agents may communicate with each other indirectly (and unbeknowingly) via the intermediary agent on the top level 2.

What reasoning paradigm is appropriate for the alarm correlation agent at the topmost level 2? Several reasoning paradigms are available, including rule-based expert systems, case-based reasoning systems, and state-transition graphs. Several commercial products that incorporate one or another of those paradigms are available.

For example, the company MicroMuse has built a product called NetCool, which is specially designed to perform the function of the topmost level-2 agent. MicroMuse has integrated NetCool with Spectrum and several other management systems. It is based on the rule-based expert system paradigm, in which a set of rules serves the function of multivendor alarm correlation, alarm triggering, and entering select data into an SLM database.

In addition, Cabletron has built a prototype system that integrates Spectrum with NerveCenter from the Seagate Corporation, where NerveCenter is the topmost level-2 agent. NerveCenter uses the state-transition graph paradigm and similarly performs interdomain alarm correlation and triggers actions based on alarms.

A clearer picture of the physical integration architecture is shown in Figure 4.13 (the SG clients have been left out). The Spectrum alarm notifier (AN) is a client daemon that receives intradomain alarms from all lower-level SSs. The AN can be configured to allow select alarms to be passed to NerveCenter (NC).

NerveCenter performs high-level reasoning over the collection of intradomain alarms, identifying any interdomain alarms. If needed, NerveCenter can communicate with other SS agents via the Spectrum command line interface (CLI). Communications can include a request of certain SSs for further bits of information, a request of certain SGs to display a warning of an eminent failure, and a request of a paging system to contact a repairperson.

The integration of Spectrum and NerveCenter at the time of this writing is in the working prototype phase and has been demonstrated at

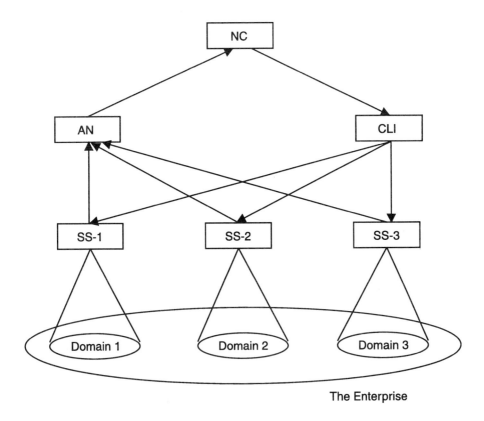

Figure 4.13 An integrated architecture with Spectrum and NerveCenter.

a few customer sites, including AT&T and MCI in the United States. Demonstration, however, is quite different from actual deployment.

One of the outstanding questions Cabletron currently is examining is the ease with which one can encode knowledge in the top-level agent (compare with question 9 in Section 4.5). In the AI community, the task is known as the problem of knowledge acquisition. For example, we may ask: How easy is it to encode new knowledge in a state-transition graph? A rule-based expert system? A case-based reasoning system?

Another important question Cabletron is studying is the ability of the topmost agent to learn and adapt itself to new problems given its experience (compare with question 10 in Section 4.5). That task is ambitious, but it takes us closer to the realization of an intelligent, self-managed enterprise management system. Cabletron has built a product named

SpectroRx that features some degree of learning and adaptability. It is an implementation of case-based reasoning. (SpectroRx is discussed in more detail in Chapter 5.)

A further consideration: BP modeling

In the final analysis, the network is just one element in a larger picture that offers an enterprise's BPs. Such BPs also depend on computer systems and applications running over the systems. When a network element is not functioning properly, fault management will help isolate it and correct the problem. But from an operations perspective, it important to understand how the business as a whole is affected.

For example, a router failure may cause a marketing forecast report to fail, or a file server crash might interrupt a nightly software distribution. We want to understand how higher-level functions can be understood and managed via the conceptual architecture illustrated in Figure 4.10.

When a standard business function such as accounts receivable is described, we discover that it is, like any BP, composed of resources and workflows. Required resource components may include client workstations, computer servers, file servers, database applications, peripherals, and the voice and data networks connecting them.

Workflows often consist of standard procedures, contingency plans, and data flows. Human operators also are involved, but because they lack a standard management interface, we ignore them here. Computing systems and network elements long have been manageable, and applications are becoming increasingly so via proprietary interfaces and current standards efforts.

We conjecture that the multilevel architecture shown in Figure 4.10 can also be applied to BP management. First, the systems, servers, applications, PBXs, peripherals, network devices, and power supplies (and operators, for that matter) can be modeled, along with the connection and dependency relationships among them. The objective, then, is to monitor the presence and health of the elements and roll the collective results up to a higher-level representation of the accounts receivable BP.

The monitoring data can be fed to state machines (or other techniques) used to model the workflows. Because the elements are distributed across multiple domains in a large enterprise, we must consider that the information will be coming from several agents and must be correlated. With the aggregate information, it becomes possible to focus on shared mission-critical resources, understand the breadth of impact when a shared element experiences troubles, and proactively manage problems on a business priority basis. This is an area of active and challenging research in the enterprise management community, and we look at it in more detail in Chapter 5.

Summary

This chapter examined SLM architecture. First we distinguished between conceptual architecture, physical architecture, and strategic architecture. We argued that it is necessary to study conceptual architecture before physical architecture and warned that it is a common error to jump into physical architectural discussions without first having examined its conceptual underpinnings.

Next, we looked at a basic conceptual architecture for SLM. We showed how SLM is built on existing enterprise management methods and argued that it is a mistake to talk about SLM without first understanding enterprise management. In addition, we showed how some management styles are network-centric, traffic-centric, systems-centric, or application-centric, but it is not hard to take the step toward a more comprehensive enterprise management system.

We embraced the view of enterprise management as cooperation and collaboration among intelligent agents, each agent having a special area of expertise. On the basis of that premise, the chapter reviewed ongoing work and achievements in AI, robotics, and data warehousing. We discussed those areas with an eye toward our enterprise management challenge.

Then we reconsidered our basic SLM conceptual architecture, incorporating insights and lessons learned from AI, robotics, and data warehousing. We described an enhanced SLM architecture that includes the notions of real-time and off-line SLM, multiple levels of abstraction, reasoning, and instruction, and alternative ways to distribute data in an enterprise management system.

Our final SLM architecture provides us with a method for situating and evaluating SLM proposals. We listed a series of 10 questions (and answers) by which we can measure the quality of SLM proposals.

Finally, the case study of Deutsche Telekom, a German telecommunications company, showed the transition of our conceptual SLM architecture to a physical embodiment.

Exercises and discussion questions

1. Early in the chapter, we said that our SLM architecture provides facilities for security control and software delivery over the enterprise. What are the requirements for security control? How can security control be embedded in the architecture shown in Figure 4.10?

2. What are the requirements for software delivery over the enterprise? How can software delivery be embedded in the architecture shown in Figure 4.10?

3. A requirement of an enterprise management system is the discovery of the components in an enterprise and their arrangement in a symbolic topological model. In the industry, that is called autodiscovery, and it is the first thing done after installation of an enterprise management system. Where do topological data fit into the architecture in Figure 4.10?

4. Suppose we have a physical embodiment of an enterprise agent and a system agent as shown in Figure 4.2. Imagine that the system agent observes a "server down" alarm, but actually a switch has failed and the server is downstream from the switch. The server is in good shape otherwise. Evaluate the following two ways of treating the problem:

   a. The system agent requests information from the enterprise agent before making a decision about the reality of the "server down" alarm.

   b. The system agent forwards the "server down" alarm to the enterprise agent, and the enterprise agent makes a decision about the reality of the "server down" alarm.

5. Beginning in 1998, the magazine *Data Communications* (McGraw-Hill) publishes a special annual advertisement section entitled "The Service Level Management Challenge." The editors describe a hypothetical but realistic enterprise with respect to its components, usage, and services. Vendors are invited to propose a physical system that can meet the SLM challenge. Each vendor is allowed two pages to describe their system and show how it satisfies the SLM requirements.

Keep in mind that the section is an advertisement, and each vendor pays up to $30,000 dollars to publish its solution. Thus, one can expect some marketing language. Of course, the pay-off is that readers will be impressed with some particular solution and contact the vendor for further information and possible procurement.

Find an issue of *Data Communications*. Study the special advertisement section and formulate answers to the following questions. Discuss your answers with your peers.

a. Pick the strongest SLM proposal. Why is it strong? Does your selection agree with the selections of your peers?
b. Pick the weakest SLM proposal. Why is it weak? Does your selection agree with the selections of your peers?
c. Pick any one proposal and grade it with respect to the 10 questions in Section 4.5. Does your evaluation generally agree with the evaluations of your peers?

Further studies

Much of the discussion in this chapter is based on advancements in DAI, robotics, and data warehousing.

Table 4.1 was adapted from early DAI studies, including Sridharan's summary paper "Workshop on Distributed AI"; Decker, Durfee, and Lesser's "Evaluation Research in Cooperative Distributed Problem Solving"; and Huhns's book *Distributed Artificial Intelligence*.

The author first proposed considerations of DAI toward an understanding of enterprise management in his 1995 paper "AI and Intelligent Networks in the 1990s and Into the 21st Century." The author's 1995 paper also discusses the relevance of advancements in robotics for enterprise management. Other good references on the topic of architectures for robot intelligence include Maes's book, *Designing Autonomous Agents: Theory and Practice from Biology to Engineering and Back,* and Minsky's classic book, *The Society of Mind.*

The section on data warehousing is based on the author's technical note "The Spectrum Data Warehouse." A very good book on data warehousing and data mining is Adriaans and Zantinge's *Data Mining.*

The references listed here are by no means exhaustive. The motivated reader is advised to perform a Web search or, better, to visit a university library to discover recent advancements in AI, robotics, and data warehousing. There is an abundance of literature on those topics, and the reader is forewarned against information overload. Nonetheless, it is likely that one will find some ideas and methods in contemporary research that will contribute to enterprise management solutions.

Earlier studies on the enterprise management architecture described in this chapter can be found in Frey and Lewis's paper "Multi-Level Reasoning for Managing Distributed Enterprises and Their Networks" and Lewis and Frey's paper, "Incorporating Business Process Management into Network and Systems Management." The case study of Deutsche Telekom is complemented by Wiepert's paper, "The Evolution of the Access Network in Germany."

For other insights in enterprise management architecture, the reader is advised to thumb through the general references listed in the Further Studies section in Chapter 1. Examples of relevant literature in that regard include Westcott's paper, "A Simple Model for Integrating Network Management," Olesen's "Network Management in Large Networks," Disabato's "Key Technologies for Integrated Network Management," Carter and Dia's "Evaluating Network Management Systems: Criteria and Observations," and Mahler's "Multi-Vendor Network Management—The Realities."

Select bibliography

Adriaans, P., and D. Zantinge. *Data Mining.* Reading, MA: Addison-Wesley, 1996.

Carter, E., and J. Dia. "Evaluating Network Management Systems: Criteria and Observations." In I. Krishnan and W. Zimmer (eds), *Integrated Network Management II.* Amsterdam, North Holland: Elsevier Science Publishers, 1991.

Decker, K., E. Durfee, and V. Lesser. "Evaluation Research in Cooperative Distributed Problem Solving." Chapter 19 in L. Gasser and M. Huhns (eds), *Distributed Artificial Intelligence,* Vol. 2. London: Morgan Kaufman, 1989.

Disabato, M. "Key Technologies for Integrated Network Management." In H-G. Hegering and Y. Yemini (eds), *Integrated Network Management III.* Amsterdam, North Holland: Elsevier Science Publishers, 1993.

Frey, J., and L. Lewis. "Multi-Level Reasoning for Managing Distributed Enterprises and Their Networks." In A. Lazar, R. Saracco, and R. Stadler (eds), *Integrated Network Management V.* Chapman and Hall, 1997.

Huhns, M. (ed). *Distributed Artificial Intelligence.* London: Morgan Kaufman, 1987.

Lewis, L. "AI and Intelligent Networks in the 1990s and Into the 21st Century." In J. Liebowitz and D. Prerau (eds), *Worldwide Intelligent Systems.* Amsterdam: IOS Press, 1995.

Lewis, L., and J. Frey. "Incorporating Business Process Management into Network and Systems Management," *Proc. 3rd International Symposium on Autonomous Decentralized Systems,* Berlin, April 1997.

Maes, P. (ed). *Designing Autonomous Agents: Theory and Practice From Biology to Engineering and Back.* Cambridge: MIT Press, 1991.

Mahler, D. *Multi-Vendor Network Management—The Realities.* In H-G. Hegering and Y. Yemini (eds), *Integrated Network Management III.* Amsterdam, North Holland: Elsevier Science Publishers, 1993.

Minsky, M. *The Society of Mind.* New York: Simon and Schuster, 1985.

Olesen, K. "Network Management in Large Networks." In D. Khakhar and V. Iverson (eds), *Information Networks and Data Communications II.* Amsterdam, North Holland: Elsevier Science Publishers, 1988.

Sridharan, N. "Workshop on Distributed AI." *AI Magazine,* Summer 1987.

Weipert, W. "The Evolution of the Access Network in Germany." *IEEE Communications Magazine,* Feb. 1994.

Westcott, J. "A Simple Model for Integrating Network Management." In D. Khakhar and V. Iverson (eds), *Information Networks and Data Communications II.* Amsterdam, North Holland: Elsevier Science Publishers, 1988.

In which we look a little deeper into some specific challenges in service level management (and enterprise management in general).

In this chapter:

▶ The event correlation problem

▶ The semantic disparity problem

▶ The component-to-service mapping problem

▶ The agent selection problem

▶ The integration problem

▶ The scaling problem

▶ The representation problem

▶ The complexity problem

▶ Case study: KLM Airlines

Special topics in SLM

This chapter discusses some particularly challenging problems in SLM. We touched on most of these problems in preceding chapters but deferred an analysis of them until this chapter.

Our first goal is simply to appreciate the problems listed to the left. If we understand a problem, then we are halfway toward a solution. In addition, this chapter describes some standard solutions that are relatively easy to implement and will get us by within a comfortable margin of acceptability.

The fact that there are indeed solutions to each problem is good news. But if we look more closely at the problems, we will see that they are good candidates for M.S. or Ph.D. theses or research papers in technical journals. Such advanced solutions are relatively harder to conceptualize and implement, but they might push the state of the science forward. Some people call this the cutting edge of science; others call it the bleeding edge.

A good example is the problem of event correlation over network, traffic, system, and application events, which we consider first. We identified the event correlation task in prior chapters but considered it more or less as a black box. Now it is time to give it full treatment, because it plays a crucial role in enterprise management.

The case study in this chapter describes an SLM experiment at KLM Airlines in the Netherlands. The experiment touches on most of the challenges discussed in the chapter.

5.1 The event correlation problem

The event correlation problem has become an issue of paramount importance in enterprise management. We will introduce the problem with a simple analogy—ourselves.

Consider that our five senses transport loads of sense impressions to our brains. Clearly, our brains cannot process all the impressions, and thus our brains have evolved such that we ignore the great majority of the impressions.

Some of the impressions, however, stand out as being significant at the moment, while others are logged into memory as being possibly significant at a later time. Think of the impressions as *events*: observations that for some reason or another simply stand out as significant and we want to hold on to them.

A hallmark of human intelligence is that we have a capacity for connecting such events. For example, we might observe that two events seem to be constantly conjoined with each other. We might hypothesize that one event is a cause of the other event or that both events are the effects of some third event yet to be observed.

Given our individual experiences, our brains have the capacity to construct an intricate network of event types that are connected by such cause-and-effect relationships (and myriad other relationships). That, of course, is what we mean when we say that we learn and grow with experience, and it captures what we mean by the terms *knowledge* and *wisdom*.

A further hallmark of human intelligence is that, given such an intricate network of events, we are able to observe a set of actual events and form hypotheses about what caused them. In addition, we are able to conjecture a set of possible events and then form hypotheses about what might happen if the events were actual. Some people argue that that is in fact the key to human survival: One's hypotheses die instead of oneself.

Now all that is fine and good and perhaps rather common sense. But the hard question is to understand how we do it. That question intrigues us, and scientists and philosophers have been trying to unravel it for a long time.

Several scientific theories have been proposed as to how the brain works. In research laboratories, those theories are constantly tested, corroborated, refuted, and modified in some way or other. That is classic science.

Some of the theories, at some stage of maturity, are carried over into industry and applied to real-world industrial problems. That is what we mean by engineering science following in the wake of pure science.

It turns out that the problem of event correlation in enterprise management is just such a case. It is not hard to see that the various monitoring agents that we have been discussing are much like a human's five senses and that event correlation over a collection of enterprise events is analogous to a human brain. Thus, we can apply theories about how the brain works to our real-world problem of event correlation in enterprise management.

Example

Consider an enterprise that is relatively simple but that nicely illustrates the event correlation challenge, which is fairly hard.

Figure 5.1 shows two networks, N1 and N2, connected by a link, L. Routers R1 and R2 host N1 and N2. There are four computer systems in the enterprise. Computer systems CS1 and CS2 reside on N1, and computers CS3 and CS4 reside on N2.

There is a client/server application, say, a database application, that is supported by the network infrastructure and the computer systems. A database server S resides on CS1, and database clients C1, C2, C3, and C4 reside on CS1, CS2, CS3, and CS4, respectively. The four client applications are GUI interfaces through which users U1, U2, U3, and U4 interact with S.

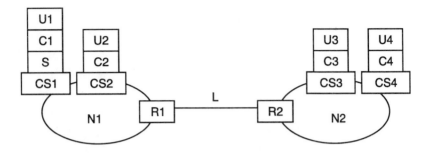

Figure 5.1 A simplistic enterprise.

Suppose we have a complete set of monitoring agents in place (Figure 5.2). A network infrastructure agent IA monitors R1 and R2. The computer system agent CSA monitors CS1, CS2, CS3, and CS4. Application agent AA monitors S, C1, C2, C3, and C4. Traffic agent TA monitors traffic that flows over N1, N2, and L. Think of a trouble-ticketing system as an agent (TTA) that monitors the users who depend on the client/server database application. The users log problems in the trouble-ticketing system when their database transactions do not work right.

Now, to appreciate the event correlation task and render it manageable and understandable, let us make three simplifying assumptions. We will lift those assumptions later in this section. As such, the way we set up the problem is a sort of straw person that we will criticize, tear apart, and reconstruct into something better. At present, however, those as-

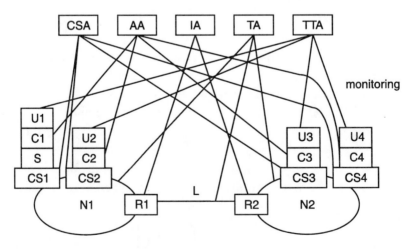

Figure 5.2 A monitoring system for the simplistic enterprise.

sumptions will help us get a good understanding of the event correlation challenge.

- ▶ Assumption 1: Each monitoring agent is good only at observing events in its domain of interest.

- ▶ Assumption 2: All events observed by all monitoring agents go into a central bucket.

- ▶ Assumption 3: No agent is aware of the events observed by its peer agents.

Clearly, we need an external event correlation agent (ECA) that reasons about the state of the enterprise based on the relationships among the components in the enterprise and the collection of events in the bucket, as shown in Figure 5.3.

For an example of the event correlation task, consider a common problem in enterprise management. Suppose the first symptoms of the problem are that U3 and U4 have logged trouble tickets indicating prohibitively sluggish behavior of their database transactions. U3 and U4 have jobs to do but they cannot get them done, and it is not their fault. The problem has been escalated to upper level management, and upper level management has phoned the enterprise overseer and demanded a fix. The business depends on it.

At this juncture, the overseer becomes a human ECA. Imagine the overseer contemplating the structure of the enterprise and the known facts (i.e., sifting through the bucket of events):

1. The TTA has deposited a series of events (i.e., trouble tickets) indicating prohibitively sluggish behavior reported by users on N2.

2. The AA has deposited a series of events in the bucket indicating connection failures issuing from C3 and C4.

3. The IA has deposited a series of events indicating resource overload on R1.

4. The TA has deposited a series of events indicating traffic overload on N1.

5. All other events are of no relevance to the problem at hand.

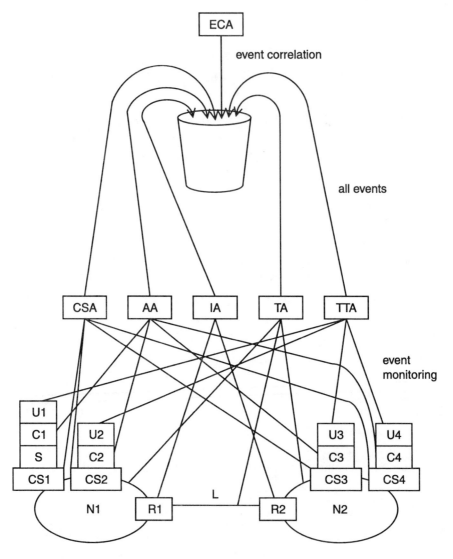

Figure 5.3 A central bucket and an event correlation agent.

The human ECA might reason thusly:

The enterprise infrastructure is clearly overstressed. First, let me determine whether any changes have been made recently regarding the structure of the enterprise. That could be the root of the problem. If not, then perhaps U3 and U4 are retrieving very large blocks of data from S simultaneously. That surely would cause the problem. I can call up U3 and U4 to check the hypothesis, or I can investigate further into traffic types and loads flowing over L, N1, and N2.

If the problem persists, there are several things I might propose to correct it:

1. Schedule the tasks of users on N1 and N2 to alleviate stress on the infrastructure.

2. Move the users on N2, who access S, to N1.

3. Move S to N2.

4. Duplicate S on N2.

5. Increase the bandwidth of N1.

6. Increase the resources of R1.

7. Add an additional network, N3, and move S and the users currently on N1 and N2 that access S to N3.

8. Redesign the client applications so that more data reside on the client, reducing the need for data traffic.

9. Some combination of 1 through 8.

The best solution, however, depends on other considerations:

If the maintenance budget is tight, then moves 1, 2, and 3 are the best bets because they are inexpensive. But I have to consider whether 1 is realistic in the long run. With 2, I might resolve the troubles of the users on N2, but I have to consider the effect of that move on the bandwidth of N1. And if I choose 3, I will have to consider whether the users remaining on N1 who have to access S might begin to experience prohibitively sluggish behavior, in which I have the same problem all over again. Is 4 a possibility? Do we have license to install a duplicate server? Would the servers have to maintain synchronization? What would be the traffic overhead of that approach?

Options 5, 6, and 7 are expensive. If I choose 5, then I have not solved the problem of resource overload on R1; if I choose 6, I have not solved the problem of traffic overload on N1. If I choose 7, then I have to worry about the effects on network performance when clients on N3 have to access other servers and applications on N1 and N2.

Is option 8 a possibility? Certainly not in the short term. But I have to find some kind of short-term solution immediately. Let me pray that the problem goes away quickly and never comes back again.

That bit of reasoning is rather dramatic, but realistic nonetheless. Most enterprise overseers can identify with it.

The task of the ECA requires natural analytical abilities, patience, and knowledge built up over many years of study and experience. An interesting aspect in the industry is that when one becomes highly competent at event correlation and enterprise troubleshooting, one is ready to move on to a higher level position. It would be wonderful if we could find some way to automate (or semi-automate) the task. Because the task requires knowledge and intelligence, we would be well advised to look to methods in AI and related disciplines to do that.

The remainder of this section looks at five reasoning paradigms that can be used to build an automated ECA and at several vendors in the industry whose event correlation products are based on those paradigms. The paradigms are:

- Rule-based reasoning (BMC Patrol, Tivoli TME, and others);

- Model-based reasoning (Cabletron Spectrum);

- State-transition graphs (SeaGate NerveCenter);

- Codebooks (SMARTS InCharge);

- Case-based reasoning (Cabletron SpectroRx).

Each method is examined along six dimensions:

- *Knowledge representation.* What is the basic structure for representing event correlation knowledge? What is the algorithm that reasons about the knowledge? How intuitive are the structure and the algorithm? Do they approximate the reasoning going on in the overseer's head?

- *Knowledge acquisition.* Assuming that we have settled on a good structure and algorithm by which to represent event correlation knowledge, how easy is it to extract knowledge from the overseer and embed it into the structure?

▶ *Computational overhead.* What is the computational expense of the automated ECA with respect to computer space and processing? What are the performance hits suffered by the enterprise infrastructure as a whole?

▶ *Scalability.* As the enterprise expands and the total number of events increase, how can we guarantee that the ECA will not buckle under with the extra load? Similarly, as we enter more knowledge into the ECA, how can we guarantee that the agent can handle the extra load?

▶ *Learning.* To what extent can we automate the knowledge acquisition task? Can we build an ECA that learns with experience as the enterprise changes?

▶ *Adaptation.* To what extent can the ECA handle problems it has never seen before? In other words, how can existing knowledge be adapted to apply to new situations?

Each dimension is important. For example, we could have an apparatus whose knowledge representation technique is intuitive, but if it is difficult to embed new knowledge in it, it is not much good. Likewise, the apparatus could be palatable with respect to both knowledge representation and knowledge acquisition, but if it is computationally expensive and does not scale, it is not much good either.

The rule-based reasoning approach to event correlation

A common approach to the event correlation task is to represent knowledge and expertise in a rule-based reasoning (RBR) system (also known as expert systems, production systems, or blackboard systems; see Figure 5.4). An RBR system consists of three basic parts: a working memory, a rule base, and a reasoning algorithm.

The working memory consists of facts. The collection of facts includes the sum total of events in the bucket and facts about the topology of the enterprise.

The rule base represents knowledge about what other facts to infer or what actions to take, given the particular facts in working memory. In the example presented earlier in Section 5.1, it is implicit that the overseer

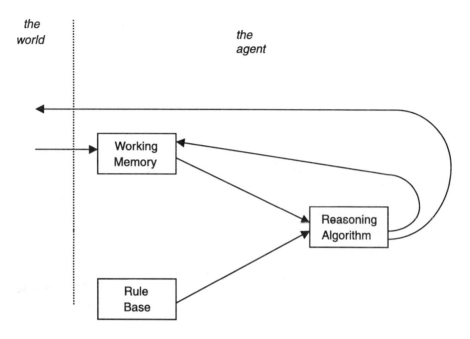

Figure 5.4   Basic structure of an RBR system.

has a set of rules in mind by which to arrive at the conclusion "the enterprise infrastructure is overstressed" and another set of rules by which to arrive at the possible options for rectifying the problem.

The reasoning algorithm (sometimes called an inference engine) is the mechanism that actually makes the inference; that operation also is implicit in the example.

The best way (albeit very simple) to think about the operation of the reasoning algorithm is to recall the classic *Modus Ponens* inference rule in elementary logic:

A	*A fact in working memory*
If A then B	*A rule in the rule base*
Therefore, B	*An inference made by the reasoning algorithm*

Because the antecedent A of the rule "If A then B" matches fact A in the working memory, we say that the rule fires and the directive B is executed. Note that B can be several kinds of directive:

▶ Add a new fact to working memory.

▶ Perform a test on some part of the enterprise and add the result to working memory.

▶ Query a database and add the result to working memory.

▶ Query an agent and add the result to working memory.

▶ Execute a control command on some enterprise component (e.g., reconfigure a router, or prohibit a certain class of traffic over a link or network).

▶ Issue an alarm via some alarm notification medium.

Regardless of the particular directive, after the reasoning algorithm makes a first pass over the working memory and the rule base, the working memory becomes enlarged with new facts. The enlargement of the working memory might be a result of the directives, or it might be a result of the monitoring agents that enter new facts in the working memory over time. In either case, on the second pass there might be other rules that fire and offer new directives and therefore new facts, and so on for each subsequent pass.

We can begin to see how we would represent the knowledge in the example. The bucket in Figure 5.3 is the working memory in Figure 5.4, and the monitoring agents in the outside world populate the memory. Now we can begin to conceptualize the kinds of rules required to represent the overseer's reasoning:

R1: *if*   load(L, t1) = high *and*
        load(N1, t1) = high *and*
        connection_failure(C3, S, t1) = true *and*
        connection_failure(C4, S, t1) = true
    then problem(t1) = overstress

R2: *if*   problem(t1) = overstress
    then add_to_memory(query(U1, t2, What are you doing?)) *and*
            add_to_memory(query(U2, t2, What are you doing?)) *and*
            add_to_memory(query(TA, t2, Show traffic by category
            from t1–10 to t1))

Rules like R1 on the first pass serve to offer a hypothesis representing the problem at hand, and rules like R2 serve to collect more information that may corroborate or dispel the hypothesis. We can imagine further rules that propose (or execute) solution options to rectify the problem. Eventually, assuming everything works right, some rule will cause the hypothesis "problem(t1) = overstress" to be retracted from working memory.

At this point, we should be able to appreciate the sort of complexity entailed by representing knowledge with RBR systems. The good news is that rules are rather intuitive, as is also the basic operation of an RBR system. The bad news is that it is not trivial to come up with a correct set of rules that behave in the way we conceptualize them. That problem shows up especially when subsequent passes over the memory and rule base issue unplanned directives or seem to be going nowhere, or when the reasoning algorithm gets caught in a nonending loop.

There are other important challenges in RBR research that we will not go into here (see the Further Studies section), for example, forward versus backward reasoning, rule ordering, rule conflict resolution, temporal reasoning, dependency-directed backtracking, and approximate reasoning.

Generally, the consensus in the industry regarding the use of RBR systems is this:

If the domain that the RBR system covers is small, nonchanging, and well understood, then it is a good idea to use the RBR method. Trying the RBR approach when those three conditions do not hold is asking for trouble.

Thus, using an RBR system to develop an automated ECA that covers the entire domain of the enterprise is not a good idea, at least in regard to the way we have set up the problem. The enterprise domain typically is large, dynamic, and hard to understand. A correct rule set at one point in time might well become invalid if the structure of the enterprise changes and the rule set is not updated accordingly. For that reason, for companies that tend to make changes frequently, RBR systems will rate poorly with respect to knowledge acquisition, scalability, learning, and adaptation. Clearly, then, it is important for a company to be aware of its expected rate of change when considering tools and techniques for an automated ECA.

But that does not mean that RBR systems do not have a place in enterprise management. For example, let us think about a single computer system as opposed to an entire enterprise. A single computer system is a much smaller entity than an enterprise. It is reasonable, then, to consider an RBR system to perform event correlation over this small domain.

In fact, that is precisely the way that most vendors who build computer monitoring agents do it, for example, BMC Patrol, Tivoli TME, Computer Associates TNG, and Platinum ServerVision. Most of those systems are one-iteration-type systems. The reasoning algorithm periodically makes a pass over the memory and the rule base and checks to see if any event (or set of events) should be escalated to an alarm. Such events include repetitious failures of logon attempts and thresholds for parameters such as disk space and CPU usage.

At this point, let us lift our first and second simplifying assumptions at the beginning of this section (namely, that monitoring agents are good *only* at observing events regarding their domains of interest and that all events from all monitoring agents go into a central bucket). That is our first step toward tearing apart our straw person and putting it back together again.

Figure 5.5 shows this new way of looking at event correlation over the enterprise, where the concepts of an event space and alarm space are made explicit. The event-to-alarm mapping function for each kind of monitoring agent is included in each agent's box, and we see now that alarms (instead of events) go into the central bucket. Several other observations regarding Figure 5.5 are in order.

First, observe that the CSA includes an RBR component that sifts through its own private bucket of events, making determinations about alarms with respect to its domain of interest. Only alarms are sent to the public bucket, while remaining events are discarded or go into an archive.

Second, observe that the external agent that sifts through the central bucket also is represented as an RBR agent, in particular an alarm correlation agent (ACA). That is a viable approach now because the number of items in the bucket should be considerably less (as opposed to a bucket of raw events). The product NetCool from Micromuse (United States) takes that approach. NetCool is a recipient of alarms from other

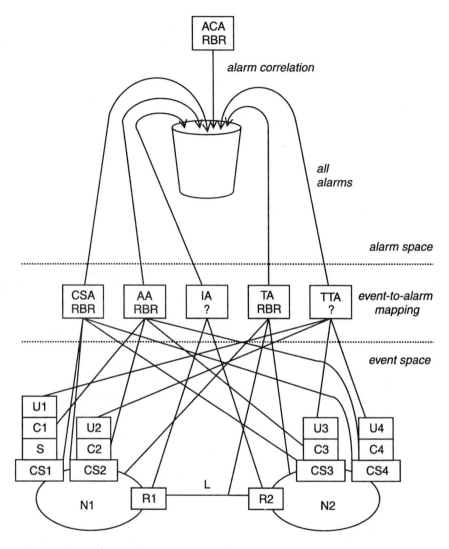

Figure 5.5 Distributing event correlation over multiple agents.

monitoring systems. If NetCool were to receive all events (instead of fewer alarms) from all monitoring agents, it would be subject to the scalability problem. Another product that uses the approach is Network Security Manager (NSM) from Intellitactics (Canada). NSM uses an RBR method to correlate (1) alarms from monitoring agents, (2) alarms issuing from intrusion detection agents, and (3) alarms issuing from biometric agents (e.g., sensors and smart cards).

Third, observe that we have distributed the event correlation task over a number of agents. It is useful to compare the approach with the discussion of DAI in Chapter 4. In general, the task is such that no one agent can perform the task alone; it requires collaboration and cooperation among a collection of agents to get it done.

Fourth, take a moment to compare Figure 5.5 with the enhanced SLM architecture in Figure 4.10. We can still have agents raising their alarms via their special mechanisms and also performing some automated control actions over the level-0 event space, all with respect to their specific domains of interest, while still passing select alarms to a higher level ACA agent that reasons about the state of the enterprise as a whole.

Finally, observe that the AA and the TA also are represented with an RBR component, but we have put a question mark in that space for the IA and the TTA. Next, we look at three candidates for implementing event correlation knowledge in the IA: model-based reasoning, state-transition graphs, and codebooks. We also examine case-based reasoning, which is a good candidate for TTA knowledge.

The model-based reasoning approach to event correlation

Consider a simple example to explain how model-based reasoning (MBR) works and how it differs from rule-based reasoning. Refer to Figure 5.6 as we work through the example.

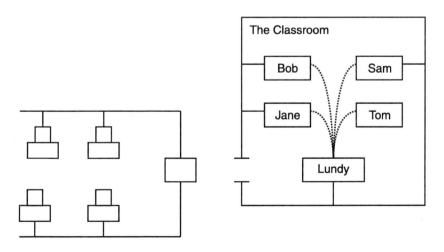

Figure 5.6 A classroom with students and a lecturer.

Suppose we have a classroom with some 20 students, and Lundy is the lecturer. Imagine that each student in the class is a model of some real computer system outside the door. Jane and Tom are models of NT servers, Bob and Sam are models of UNIX workstations, and so forth. Now imagine that the Jane, Tom, Bob, and Sam models have a way to communicate with their real-world counterparts. They ping their counterparts every 10 minutes to make sure they are up and running.

The classroom is a model of a subnet. Lundy, the lecturer, is a model of the router that hosts the subnet, and Lundy is in communication with the router. There is a real-world system outside the door, and the classroom is a representation of that system.

Lundy and the students are happily pinging their real-world counterparts when Jane suddenly pings her NT server and does not get a response. After two more pings with no response, Jane sends a message to Lundy asking if he has heard from his router. If Lundy answers yes, Jane infers that there is a fault with her NT server and raises an alarm accordingly. But if Lundy answers no, Jane reasons that probably her NT server is in good shape and the real fault is with the router.

Suppose Lundy does answer no. Then Jane suppresses her alarm, and Lundy, not getting a response from his router, quickly raises an alarm. Jane suppresses her alarm because she knows how she stands in relation to the whole system: (1) her counterpart is attached to the subnet, (2) Lundy's counterpart hosts the subnet, (3) her state depends (in part) on the state of Lundy, but (4) Lundy is in bad health.

The thrust of the example is that event correlation is a collaborative effort among virtual intelligent models, where the models are software representations of real entities in the enterprise.

At this point, we will digress for a moment. The astute reader probably will be carried back to our discussion in Chapter 4 of DAI and object-oriented analysis (OOA) and rightly be a bit puzzled. It would seem that MBR concepts are similar to DAI and OOA concepts. For example, a *model* in MBR seems to be much the same as an *agent* in DAI and much the same as an *object* in OOA.

That reader would be quite on the mark. But why do we have three different names (MBR, DAI, and OOA) referring to roughly the same thing? The best answer is that three different communities happened to be on the same wavelength at roughly the same point in history (the early

1980s). Those communities are still strong and evolving as more research and practical results bear on their goals.

That is the sort of thing academic symposia are made of. In the professional literature, one often sees a "Call for Participation," in which it is noted that researchers in different communities are working on similar problems, and the goal of the symposium is to get these researchers together in hopes of cross-fertilization and advancement of science. Sometimes it works, sometimes it does not. Regardless, it is classic science.

Now let us continue with our discussion. We describe an MBR system for event correlation over the enterprise as follows:

▶ An MBR system represents each component in the enterprise as a model.

▶ A model is either (1) a representation of a physical entity (e.g., a hub, router, switch, port, computer system) or (2) a logical entity (e.g., local, metropolitan, or wide area network (LAN, MAN, WAN); domain; service; BP).

▶ A model that is a representation of a physical entity is in direct communication with the entity it represents (e.g., via SNMP).

▶ A description of a model includes three categories of information: attributes, relations to other models, and behaviors. Examples of attributes for device models are *IP address*, *MAC address*, and *alarm status*. Examples of relations among device models are *connected to*, *depends on*, *is a kind of*, and *is a part of*. An example of a behavior is *If I am a server model and I get no response from my real-world counter part after three tries, then I request status from the model to which I am connected and then make a determination about the value of my alarm status attribute.*

▶ Event correlation is a result of collaboration among models (i.e., a result of the collective behaviors of all models).

The best example in the industry of the MBR approach is Spectrum from Cabletron Systems. Spectrum contains model types (known as classes in OO terminology) for roughly a thousand kinds of physical and logical entities, where each model type contains generic attributes, relations, and behaviors that instances of the type would exhibit.

The first thing one does after installing Spectrum is to run the Spectrum autodiscovery. Autodiscovery discovers the entities in the enterprise and then fills in the generic characteristics of each model with actual data. As monitoring happens in real time, the models collaborate with respect to their predefined behaviors to realize the event correlation task.

Let us see how the MBR approach used in Spectrum compares with the six dimensions described on pages 164 and 165.

With respect to the dimension of knowledge representation, we have a natural match between the MBR paradigm and the structure of real enterprise systems; thus it follows that we have a good match with the way an enterprise overseer thinks about the enterprise. That is much better than a collection of rules. Although experts appear to use rules, they think of an enterprise first in terms of its components and structure.

With respect to knowledge acquisition, the task is to define the structure of a model with respect to its attributes, relations to other models, and behaviors. At first that would seem to be a formidable task, but it is not so bad. Consider that Spectrum contains generic model types for a large number of enterprise entities. After running autodiscovery over the enterprise, a subset of those models is instantiated with relevant attributes, relations, and behaviors. Therefore, the knowledge acquisition effort is reduced at the outset.

What happens if there exists an entity in the enterprise for which a model type is not available? There are two ways to approach that situation. First, one can exploit the "is a kind of" relation to embed a new model type in the existing model type hierarchy. In OO terminology, that relation is called inheritance, and the model type hierarchy is analogous to a class hierarchy.

Suppose we have a generic model of a router replete with placeholders for attributes, relations, and behaviors that most routers share. We also have a set of vendor-specific router model types, each being a kind of generic router. Now suppose a vendor produces a new and improved router (say, a DifServ router) and we want to manage it. Then we can derive a new model type from the generic router model type. The derivative model inherits the characteristics of its parent. One also can add new characteristics to the derivative model to distinguish it from its siblings.

The second way is to implement new model types in C++ code and link it with the existing model type hierarchy, a method that requires an experienced software engineer.

Consider this example. The next generation Internet (Internet2) is currently in the research and development stage. If we were to explore how Internet2 could be managed with Spectrum's MBR approach, we first would examine the new sorts of devices and mechanisms that support Internet2. The first choice is to look for similar existing model types from which to derive new management models. If that does not work, we have to build new model types in C++ from scratch. In fact, such exploration is under way as of this writing.

In very large domains, computational overhead and scalability can pose a problem even when all necessary model types exist. The usual way around that problem is to assign enterprise management agents to individual domains, where domains may be geographical or logical partitions of the enterprise. Another way to alleviate the problem is to configure models so they communicate via traps that issue from their real counterparts, as opposed to pinging them periodically. (Those methods were discussed in some detail in the case study of Deutsche Telekom in Chapter 4.)

Now, creating mechanisms for learning and adaptability is notoriously hard. Ideally, each model type would be able to adapt its behavior over time and with experience. Further, the nature of collaboration among multiple models would evolve as new alarm scenarios are faced and resolved. That is a bit much to ask given the current state of the science, although there is heavy research in the area. Later in this section, we discuss an approach to the learning/adaptation challenge that holds promise: case-based reasoning.

A simple form of adaptability is available with Spectrum's MBR approach. Spectrum has a background autodiscovery agent that continuously watches for additions of new components in the enterprise. When a new component is detected, Spectrum incorporates a model of the component into the overall enterprise structure and informs an administrator accordingly.

We have seen that event correlation using Spectrum's MBR approach is achieved by a collaboration of models. It is interesting to compare an alternative, optional way to implement event correlation in Spectrum using a product called SpectroWatch. SpectroWatch is a classic RBR system. SpectroWatch can be used to formulate rules that describe how events are mapped into alarms. The advantage of that optional approach is that it is easy to do, because a GUI guides you through the process and C++ programming is not required. The disadvantage is that it may suffer the usual deficiencies of RBR systems: If the domain is large and thus the number of rules is large, the performance of event correlation can be jeopardized. That observation echoes our principle that the RBR approach can be useful if the domain of coverage is kept small.

An example of a hybrid RBR/MBR system in the telecommunications space is NetExpert developed by OSI in the United States. NetExpert uses classes, objects, attributes, and relationships to represent network entities, but uses a rule-based engine to conduct intelligent analysis.

The MBR approach has met with a fair amount of success in the industry. Thus, in Figure 5.5, we can replace the question mark in the IA box with MBR. However, MBR is not the only approach to event correlation. The following sections describe two other popular approaches: state-transition graphs and codebooks. The former method is the foundation of the product NerveCenter from the Seagate Corporation, and the latter method is the foundation of the product InCharge from SMARTS.

The state-transition graph approach to event correlation

We have made a distinction between events and alarms. Our goal is to reduce a multitude of events into a few significant alarms, and the mechanism that does that is event correlation.

Thus far, we have looked at two approaches to event correlation: the RBR approach and the MBR approach. Let us see now how the state-transition graph (STG) approach differs from those approaches.

The key concepts in the STG approach are a token, a state, an arc, a movement of a token from one state to another state via an arc, and an action that is triggered when a token enters a state.

To see how the STG apparatus works, consider the classroom scenario presented on page 172. There is a subnet hosted by a router, and the subnet contained several UNIX workstations and NT servers. One of the NT servers (Jane's server) failed, but the real reason the NT server failed was because the router (Lundy's router) failed.

We want the ECA to reason that Jane's NT server is only apparently at fault but Lundy's router is the real fault. Figure 5.7 shows an STG that can reason through that.

A first failure of Jane's NT server causes a token to move into a ground state. An action of the ground state is to ping Jane's NT server after 1 minute elapses. If the ping does not show that Jane's NT server is alive, the token moves to the yellow-alarm state; otherwise, the token falls off the board. An action of yellow alarm is to ping Jane's NT server a second time. If the action returns no, the token moves to the orange-alarm state, where the action is to ping Lundy's router. Depending on the state of Lundy's router, the token moves to either the no-alarm state or the red-alarm state and the appropriate action is taken accordingly.

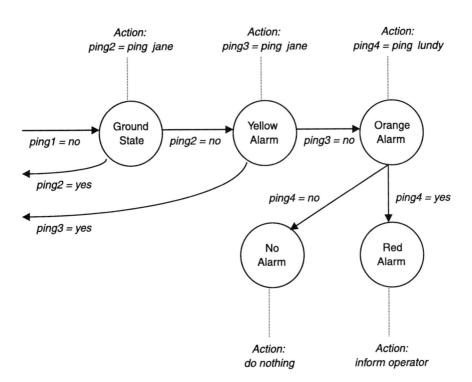

Figure 5.7 A sample state-transition graph.

Note that the STG in Figure 5.7 covers a single domain of interest, namely, Jane's NT server. An enterprise, however, might contain thousands of diverse kinds of components; thus, it might seem that a large number of STGs may be required to cover the enterprise. Fortunately, because a generic STG can apply to the same component type, there is no need to build separate STGs for like components. That considerably reduces the number of STGs required to manage an enterprise.

With the STG approach, it is possible to instrument STGs so that they confer with other STGs instead of a device. For example, the action of the orange-alarm state in Figure 5.7 is to ping Lundy's router. Instead, it could have been to ping a separate STG that covers the router. Note that this sort of design comes close to the collaboration of virtual models described in the earlier section on the MBR approach.

The best example in the industry of the STG approach is SeaGate's NerveCenter. NerveCenter typically is integrated with HP OpenView, although it can operate in standalone mode by communicating directly with managed devices via SNMP.

HP OpenView is a good example of an enterprise agent that performs event monitoring only. Thus, NeverCenter adds extra value to OpenView with its event correlation mechanism. Figure 5.8 shows the structure of the OpenView/NerveCenter integration.

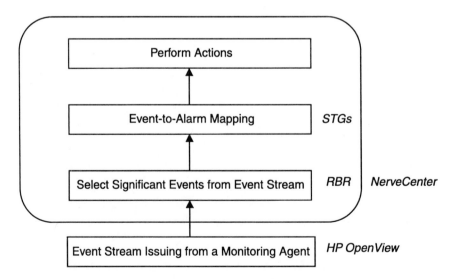

Figure 5.8   The structure of the OpenView/NerveCenter integration.

The first thing to notice in Figure 5.8 is that NerveCenter uses the classic RBR method to select significant events from an event stream, and only those events are passed to a set of STGs to perform the event-to-alarm mapping function. In the commercial literature, one often sees NerveCenter described as a rule-based system, which is somewhat misleading. NerveCenter uses two kinds of methods in tandem: RBR and STGs.

Now, how does the STG approach compare with the six dimensions described on pages 164 and 165? The knowledge representations (RBR and STGs) are fairly intuitive. However, the difficulty of the knowledge acquisition task is rather high because we have to construct both rules and STGs and make sure they match up. Further, if we are not careful in constructing a network of collaborating STGs, we can end up with the same sort of unexpected behavior that is well known to be a drawback of pure RBR approaches.

Computational overhead and scalability, as with any system that becomes too large, can pose a threat. Obviously, learning and adaptability are not inherent features of the STG approach.

Nonetheless, the STG approach taken in NerveCenter has met with a fair amount of success in the industry. If we heed the general admonishment of keeping the domain of coverage small and manageable, then we are more likely to build a useful event correlation system with the STG approach.

Thus, we have a second candidate for the question mark in the IA agent box in Figure 5.5: RBR/STG. The next section describes a fourth candidate, the codebook approach.

The codebook approach to event correlation

The major concepts in the codebook approach to event correlation are a correlation matrix, coding, a codebook, and decoding. To see how the codebook apparatus works, let us look at a very simple example.

Consider a small domain of interest in which there are four events (e1, e2, e3, and e4) and two kinds of alarms (A1 and A2). Now suppose we know the sets of events that cause the alarms. We can organize the information in a correlation matrix as follows:

	A1	A2
e1	1	1
e2	0	0
e3	0	1
e4	0	0

The matrix tells us that an occurrence of event e1 indicates A1, whereas a joint occurrence of e1 and e3 indicates A2.

Coding transforms the matrix into a compressed matrix, called a codebook. It is not hard to see that the preceding matrix can be compressed into a simpler codebook:

	A1	A2
e1	1	1
e3	0	1

The codebook tells us that only e1 and e3 are required to determine whether we have A1, A2, or neither. With the compressed codebook, the system is ready to perform event-to-alarm mapping. That is called decoding. The system simply reads events off an event stream and compares them with the codebook to infer alarms.

We could also perform the event-to-alarm mapping function with the original correlation matrix as well as the compressed codebook. However, it is not hard to appreciate the gain in speed when we use the latter. The codebook tells us to be on the lookout for two events instead of four events, and there is considerably less codebook lookup compared with the original matrix.

Second, the codebook approach allows us to compress a very large correlation matrix, which is hard for the human mind to comprehend, into codebooks that are more comprehensible. To see that, consider the correlation matrix and the two derivative codebooks in Figure 5.9. We will not explain here how the coding algorithm works (see the Further Studies section). The reader will have to trust that the two codebooks are correct.

The first codebook collapses the correlation matrix into the three events e1, e2, and e4. The codebook can distinguish among all six alarms; in some cases, however, it can guarantee distinction only by a single

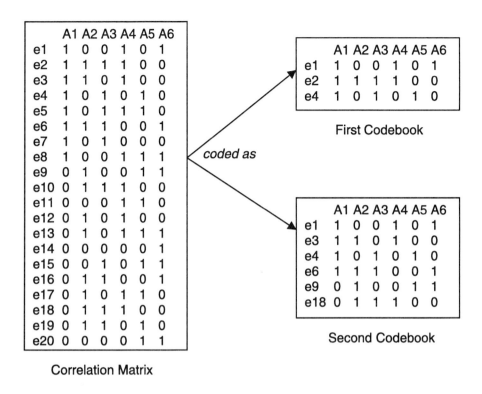

Correlation Matrix

Figure 5.9 A correlation matrix and two derivative codebooks.

event. For example, A2 and A3 are distinguished by e4, but a lost or spurious generation of e4 will result in a potential decoding error (i.e., incorrect event-to-alarm mapping).

The second codebook resolves that problem by considering six events: e1, e3, e4, e6, e9, and e18. The second codebook is such that a loss or spurious generation of any two events can be detected and any single-event error can be corrected.

We have to pay careful attention to a few things in the codebook approach to event correlation. First, how do we come up with the original correlation matrix? If the original matrix is correct, then we have an opportunity for compressing the matrix into a codebook that hides noise and irrelevant events. If it is not correct, we run into the familiar problem of garbage-in garbage-out, which is reminiscent of the classic knowledge acquisition problem.

Assume that the original matrix is indeed correct. Can we be assured that a high-performance codebook will be generated? Consider the following simple example:

	A1	A2
e1	0	1
e2	1	0

A codebook for that matrix will be equal to the matrix. In general, the compression factor of a correlation matrix depends on the patterns of data collected in it. Depending on the patterns, the compression factor may be anywhere from very high to very low.

Finally, how do we choose from multiple codebooks? In general, given any correlation matrix of reasonable size, there will be a large number of possible codebooks. For example, the discussion of Figure 5.9 showed that the simplest codebook may not be resilient against errors. On the other hand, a codebook that is resilient to errors may sacrifice understandability.

The best example in the industry of the codebook approach is In-Charge, developed by System Management Arts (SMARTS). InCharge typically is integrated with either HP OpenView or IBM NetView.

InCharge also includes an event-modeling language based on classic OO techniques, including class/subclass development, inheritance, and event definition. For example, a class TCPPort and a class UDPPort may inherit the general attributes of a class Port. However, the event Packet-LossHigh for a TCPPort will have a different definition than Packet-LossHigh for a UDPPort.

InCharge contains a generic library of networking classes. Given those classes, we can derive domain-specific classes by adding the appropriate attribute and instrumentation statements to produce an accurate model of the domain. Finally, we add event definitions to the model. That approach comes close to the MBR approach.

The knowledge representation technique of the codebook approach is intuitive. We build a correlation matrix, then a coding algorithm transforms it into a codebook. Event-to-alarm mapping is achieved by comparing incoming events with the codebook. Knowledge acquisition is achieved by expanding the correlation matrix with new knowledge and then rerunning the coding algorithm.

If we pay attention to the potential problems, the codebook approach can be efficient. Particularly, we have to make sure that the original correlation matrix is accurate.

The case-based reasoning approach to event correlation

So far, we have examined four approaches in the industry for performing event correlation over the enterprise: RBR systems, MBR systems, STGs, and codebooks. Each of those methods represents knowledge and reasoning in different ways. One thing they have in common is that they rate rather low with respect to learning and adaptability. The final method, case-based reasoning, shows promise in that regard.

Recall that the outstanding problem with the RBR approach is maintenance. If an RBR system covers a domain that remains relatively constant, a correct system needs little maintenance. If, however, the system covers unpredictable or rapidly changing domains, two problems inevitably occur: (1) the system suffers the problem of brittleness and (2) the development process suffers the problem of knowledge acquisition bottleneck.

Brittleness means that the system fails when it is presented with a novel problem. The counterpart of the brittleness problem is the system's lack of ability to adapt existing knowledge to a novel situation and learn from experience.

A knowledge acquisition bottleneck can result when a knowledge engineer tries to devise special rules and control procedures to cover unforeseen situations. In that case, the system typically becomes unwieldy, unpredictable, and unmaintainable. With rapidly changing domains, the system can become obsolete quickly. The options are to limit the coverage of the RBR system or to search for other approaches.

An alternative approach toward alleviating those problems is to represent the requisite knowledge in a case-based reasoning (CBR) framework. CBR offers potential solutions to the problems of brittleness and knowledge acquisition bottleneck. The goals of CBR systems are to learn from experience, to offer solutions to novel problems based on experience, and to avoid extensive maintenance.

The basic idea of a CBR is to recall, adapt, and execute episodes of former problem solving in an attempt to deal with a current problem.

Former episodes of problem solving are represented as cases in a case library. When confronted with a new problem, a CBR system retrieves a similar case and tries to adapt the case in an attempt to solve the outstanding problem. The experience with the proposed solution is embedded in the library for future reference. The general CBR architecture is shown in Figure 5.10.

A challenge of the CBR approach is to develop a similarity metric so useful cases can be retrieved from a case library. We would not want the system to retrieve the case that simply has the largest number of matches with the fields in the outstanding case. Some of the fields in a case are likely to be irrelevant and thus misguide the system. Relevance criteria arc needed to indicate what kinds of information to consider given any particular problem at hand.

An example of a relevance criterion is the following: The solution to the problem "response time is unacceptable" is relevant to bandwidth, network load, packet collision rate, and packet deferment rate.

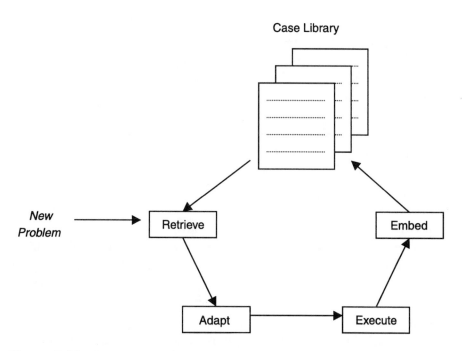

Figure 5.10   The general CBR architecture.

Note that relevance criteria are not the same as rules in an RBR system. Relevance criteria simply tell the system what cases to look at; they do not tell the system what to do with those cases.

How do we acquire relevance criteria? Current research involves the application of machine-learning algorithms such as neural networks and induction-based algorithms to an existing case library. The output of the algorithm is a list of relevance rules. For the present, however, a pragmatic approach is to handcraft and test them manually.

An additional challenge for the CBR approach is to develop adaptation techniques by which the system can tweak an old solution to fit a new problem, although the new problem might not be exactly like the old problem. Several kinds of adaptation techniques exist, some of which are illustrated as follows.

Consider the outstanding problem "response time is unacceptable" and imagine that only one source case is retrieved from the case library, as shown below. In this example, the resolution is "page_space_increase = A," where A is a value that indicates the amount by which to increase the page space of a server, determined by the function f.

> Problem: response time = F
> Solution: A = f(F), page_space_increase = A
> Solution status: good

This adaptation method is called *parameterized adaptation*. It is a method for adjusting the solution variable of an outstanding problem relative to the problem variable, based on the relation between the solution and problem variables in a source case. Everything else being equal, the outstanding problem "response time = F*" should propose the solution "page_space_increase = A*," where F* and A* stand in the same relation as F and A in the source case. The proposed solution in the outstanding case, therefore, would look like this:

> Problem: response time = F*
> Solution: A* = f(F*), page_space_increase = A*
> Solution status: ?

How does one acquire functions like f? The safest method is to handcraft and test them, but there are several other ways to represent the f function.

The simplest is a lookup table, where values of A not in the table are calculated by interpolation. Also, learning f from existing data in the case library can be looked on as a function approximation problem and thus lends itself to neural network methods that are generally good at function approximation, for example, counterpropagation and back-propagation.

Note also that f does not have to be a function per se. For other kinds of problems, f might be a sequence of steps or a decision tree. Suppose a retrieved case holds a simple procedure as follows:

Solution: reboot(device = client1)

where *reboot* is a process and *client1* is the value of the variable *device*. Suppose this case is just like an outstanding case, except that in the outstanding case the value of *device* is *server1*. Thus, the advised solution is

Solution: reboot(device = server1).

This method is aptly called *adaptation by substitution*.

Now, because it is impolite to reboot a server without warning, we might want to prefix the step "issue warning to clients" to the advised solution and enter the case back into to the case library. In future instances of similar problems, the preferred procedure will be regulated by the value of *device*.

In this example, the proposed solution was adapted manually by a user, so this technique is called *critic-based adaptation*.

There are several generic CBR systems in the industry, for example, CBR Express from the Inference Corporation and SpectroRx from Cabletron Systems.

SpectroRx is an add-on application for Spectrum. As described earlier, Spectrum performs the event correlation task using the MBR method. Once a fault is identified, however, there remains the problem of finding a repair for the fault. Clearly, experience with similar faults is important, and that is just the kind of knowledge that SpectroRx allows us to develop.

An interesting anecdote regarding SpectroRx is as follows. Version I of SpectroRx was shipped in 1995 with an empty case base, sometimes called in the industry a knowledge shell. The user was expected to build an initial seed case library manually, after which the system would expand and become increasingly fine-tuned with use.

The problem was that users were not too keen on the idea of first having to build a seed case library, although they liked the general idea of CBR. Many requests came in to ship SpectroRx with a generic seed case library. But now the engineers who developed SpectroRx had a problem. How could one build a generic case library, when any two enterprises are likely to be quite different, each having different components, services, and configurations?

After much research to try to understand what a generic case library would look like and to determine whether there should be just one or several generic case libraries, an engineer proposed a simple, ingenious solution: Suppose we take the event-to-alarm mapping knowledge in Spectrum and transform it into a set of cases?

Part of the beauty of that solution is that although Spectrum contains more than a thousand models for popular enterprise components, it is never the case that an enterprise will be composed of each kind of component. After running Spectrum Auto-Discovery, only one to two hundred models are actually instantiated. So, the seed case library derives from the event-to-alarm mapping knowledge that is related to only those instantiated models.

Version II of SpectroRx was shipped in 1996 with a case library that holds just one case, which says:

Problem: Your case library is empty.
Solution: Press the Execute Solution button and we will build a
          seed case library for you in about 10 minutes.
Solution Status: ?

That solution solves the problem elegantly. For any particular enterprise, the seed case library will be especially tailored to reflect the components, services, and configurations contained in it.

In sum, the CBR reasoning mechanism complements an IA agent, and we have shown how it complements the Spectrum MBR reasoning mechanism. Note that CBR would not be a good reasoning mechanism for an IA agent exclusively because the ontology of CBR does not lend itself to a representation of the structure of an enterprise.

Look back at Figure 5.5. A CBR-like agent also would seem to be appropriate for the TTA agent. For example, it is arguable that the structure of a case is much like the structure of a trouble ticket, and a case library is much like a trouble-ticket database. In addition, a CBR-like agent would seem to be a viable option for representing the reasoning mechanism for ACAs, CSAs, and AAs, although no one has tried it to date.

More on distributed event correlation

In this section, we raise the bar on event correlation a bit higher. Look at Figure 5.5 again. Various monitoring agents are receiving events from their respective domains of interest, mapping the events to alarms, and passing alarms to a central alarm bucket. For the sake of the discussion, in this section we do not worry about how the agents implement the event-to-alarm mapping function. Just assume they do it.

Now let us finally raise the third simplifying assumption stated on page 161, that no agent is aware of the events observed by its peer agents. For example, it would seem that in some cases the AA would benefit from an awareness of the observations of the IA.

Suppose the AA observes a connection failure when client C3 requests data from S. The AA raises an alarm accordingly. If there are a hundred such failures on clients C3, C4, ..., C103 (all of which reside on N2), then the AA likewise raises a hundred alarms.

Given the structure of the simple enterprise in Figure 5.5, we human experts would reason that there really is nothing wrong with C3, C4, ..., C103. There is no mention of problems with C1, C2, or S. Therefore, there must be a problem with N2, R2, L, or R1. But the AA does not know that. The AA is not as smart as we are. Clearly, if the AA was aware that the IA had raised a single alarm on R2, the AA could reason that it should suppress its alarms on C3, C4, ..., C103.

What is called for, then, is a method of communication between the AA and the IA. We can generalize that idea so that in principle each monitoring agent can request and receive event information from its peers.

Look at Figure 5.11. For purposes of illustration, we have recast the layer of monitoring agents in Figure 5.5 as a circle of communicating agents, much like a round-table discussion. To make the problem more

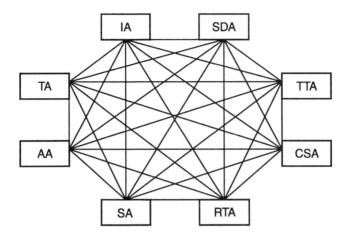

Figure 5.11 A distributed, intercommunicative management system.

interesting, we have included a special-purpose agent that measures the application response time (RTA) and two other agents we have not talked about so far in this book: a software distribution agent (SDA) and a security agent (SA).

The lines in Figure 5.11 should be understood to mean "can communicate with." The management system is fully connected, so in theory each agent can communicate with any peer agent. Here we are not concerned about the mechanism by which they communicate. (We will worry about that in Section 5.5, in discussion of the integration problem, and some more in Chapter 6.) For now, let us worry only about the content of a communication. Under what circumstances should two agents exchange alarm information?

Consider the responsibilities of an SDA. For a Web server farm consisting of hundreds of NT or UNIX servers, it would be expensive to replace or upgrade the operation systems and the applications on each server every time a vendor came out with a newer version. It would be preferrable that an agent do that automatically, which is the responsibility of the SDA. Good examples in the industry are Novadigm NDS, Metrix WinWatch, and MicroSoft SMS.

Suppose the SDA is in the middle of a large software distribution session over a server farm and a router fails. The SDA raises an alarm about unfinished business and simply stops. The manager of the farm then has to correct the problem and restart the software distribution session from scratch. If the session requires a full day to complete, then precious time and work are wasted.

But suppose an IA can detect (or predict) a router failure before it has an effect on software distribution. The IA can be instrumented to send a message to the SDA telling it to suspend work until further notice. Then, when the router comes back on-line, the IA sends a second message to the SDA telling it to continue where it left off.

Take another example, this time regarding communication with the IA, CSA, and TA. Suppose the SDA is ready to initiate a software distribution session. The SDA may send a message to the IA, CSA, and TA asking whether there is any reason not to proceed. If no agent is aware of any alarms on any components on which the distribution depends, then the SDA starts the session. Otherwise, the SDA waits an hour and asks the same question again.

Recall the discussion of MBR. With the MBR approach, models of enterprise components confer with each other to perform event correlation. The same thing is happening in our discussion here, but at a higher level of communication. In Spectrum, for example, all the models "live" inside a single software application, but here the models (i.e., applications) coexist and live in a larger system, likely to be distributed over the enterprise. The notion of reasoning about the state of the enterprise via collaboration is the same in both cases.

5.2   The semantic disparity problem

In its simplest form, the semantic disparity problem is this: The ways that ordinary users, business executives, and computer scientists think about information technology (IT) are different.

Consider intercontinental e-mail in support of a global business. A user of e-mail knows only when e-mail is working right or not working right. It is not unusual for a user to find out that an e-mail sent a week ago has not arrived. That is frustrating and typically causes the user to distrust e-mail. Sometimes a user will begin to depend on e-mail accompanied by a backup phone call, which is clearly time-consuming, expensive, and undesirable.

A business executive sees intercontinental e-mail as a way to increase the productivity of the company as a whole. It is well known that communication is proportional to productivity. If the executive finds out that e-mail is unreliable, the first thought is how that affects the goals of the business; the second thought is how that affects the morale and well-being of the users who depend on e-mail to do their jobs. An executive who learns that users always back up their e-mails with phone calls rightly becomes infuriated.

Computer scientists think of e-mail in terms of the underlying components that support it, for example, computer configurations, e-mail server configurations, client e-mail applications, managing the Internet, protocols, and the like. If e-mail becomes dysfunctional, the computer scientist's thoughts tend toward those items.

In sum, the meaning (i.e., semantics) of *intercontinental e-mail* and other IT-related terms is different depending on the point of view. Because the interests, language, and conceptual frameworks of users, business executives, and computer scientists are divergent, it is a bit of a strain for them to communicate about IT.

The concept of a *service* is one way to bridge the semantic gap among different mindsets. For example, step 2 of the SLM methodology presented in Chapter 3 urged that services be named and described with simple, common sense language. Step 3 urged that service parameters and service levels likewise be named and described with simple language. The names and descriptions should be expressed without regard to technical details; rather, they should be expressed with respect to the user's point of view and in the user's language. Thus, after steps 2 and 3 have been completed, the user should be satisfied and content to exit SLM discussions without further worries.

The business executive certainly is concerned with user happiness and morale but also with the business as a whole and how the identified services contribute to the goals of the business. Assuming that one has followed the SLM methodology properly, step 1 will have identified the business goals, step 2 will have identified the services on which the business goals depend, and step 3 will have identified service levels. Thus, the executive should be satisfied also and be content to exit SLM discussions without further worries.

Unfortunately, the computer scientist is left holding the ball. Users and executives are convinced that all things are in right order, but the scientist has plenty of worries. Services and service parameters have been couched in rather subjective language, and the scientist now has the problem of translating that subjective language into something objective and measurable.

We see then that the SLM methodology solves part of the semantic disparity problem because it reconciles IT points of view among users, executives, and computer scientists with the concept of service. However, it introduces another manifestation of the problem. How does the computer scientist reconcile the subjective concepts of *service* and *service parameter* into something that is objective and measurable?

There are three approaches to the second manifestation of the semantic disparity problem: the user-centric approach, the happy-medium approach, and the techno-centric approach.

- In the user-centric approach, the computer scientist tries to find some objective way to measure the services and service parameters identified in steps 2 and 3 of the SLM methodology.

- In the happy-medium approach, the scientist tries to influence the output of steps 2 and 3 in the SLM methodology so that the selected services and service parameters are both meaningful to the user and objectively measurable. Note that this approach comes close to a violation of the principle to disregard technical details when establishing SLM requirements.

- In the techno-centric approach, the scientist tries to convince users and executives that objective measurements of low-level component parameters do indeed reflect the health of the higher level services and service parameters identified in steps 2 and 3 of the SLM methodology. In other words, component parameters can be mapped into higher-level service parameters.

Suppose we have a distributed service, say, "cooperative proposal writing and pricing," that depends on a database server, a dozen users who perform specialized transactions over the database, and a distributed document-handling application. Further assume that the following five parameters have been identified for this service. (These parameters apply to most distributed services.)

▶ Availability;

▶ Response time;

▶ Jitter;

▶ Data integrity;

▶ Data security.

It is a worthwhile exercise to examine each parameter with respect to the three approaches. Following the SE practices discussed in Chapter 3, a good scientist would always investigate the user-centric approach first and to the fullest before proceeding to the other two approaches.

For example, what does "availability" mean to the user? If this question was posed to a group of users, the consensus probably would be that they want their tools of trade to always work and not surprise them. They do not want to try a routine transaction on a database that worked fine last week only to see a weird message pop up on the screen. That is disrupting to both their states of mind and their jobs if they have tasks that must be finished by the end of the day. How can the scientist objectively measure this sense of availability?

The user-centric approach, taken to the extreme, suggests that the scientist finds some way to test all transaction types before they are actually executed by the user. Clearly, that is not a good idea. It would likely clog up the system and degrade the service.

The happy-medium approach suggests that the scientist find some way to execute generic, periodic database transactions that are generally representative of the sum total of user transaction types. This is a better idea, but now the scientist has more thinking to do. What would such a generic transaction look like? First, if client transactions issue from multiple geographical locations, it is not likely that one generic transaction would be sufficient. If multiple domains are involved, the solution calls for multiple generic transactions that cover geographically dispersed users. Second, if client transactions are of a different type (e.g., database query versus entry, or retrieval of very large chunks of data versus small chunks), it is questionable whether one generic transaction would be sufficient.

The good news is that the question of what constitutes a set of generic transactions is not insurmountable. A good scientist is intrigued by such questions and should be able to articulate and provide a good argument for some particular set of generic transactions that represent the health of the service.

The less-than-good news is that we actually have to do it. Following the SE practices discussed in Chapter 3, that is the problem of transforming an analysis model into a design model. We have to find commercial monitoring agents that can be configured to issue such transactions, notify us if the transaction is unsuccessful, and record the results in an off-line database for later SLM reporting. Failing that, we have to build such a system ourselves. Either way, however, we know what needs to be done. Section 5.4 takes up this topic in the discussion of the agent selection problem.

The astute scientist will not forget about users' "no surprises" requirement. Recall that part of a user's concept of availability is to reduce the element of surprise. Fortunately, if the scientist can determine whether a tool of trade is unavailable by issuing a set of generic transactions, it is a small step to make that information available to users.

Imagine a screen like the one shown in Figure 5.12 that is accessible to a large group of users via public electronic display or a Web browser. The screen shows which services are accessible by which groups of users and the expected time to repair. Suppose someone from the United States is in Argentina on a business trip and cannot get access to some data back in the United States. That person could go to a Web site to see who can access the data and ask that other person to send the data across the Internet via e-mail.

Response time, jitter (defined as the variation over response time), and data integrity are straightforward extensions of our treatment of availability. If we can build an agent that performs a set of generic transactions that represent the service's availability from the user's point of view, then with a little more work we also may build the agent such that response time, jitter, and data integrity also are measured.

Now let us consider the techno-centric approach to availability. There is a sense of "availability" that can be measured by monitoring the underlying servers, network devices, transmission media, and computer systems on which the "cooperative proposal writing and pricing" service depends. We consider the techno-centric approach more fully in Section 5.3. First, let us provide a simple analogy that puts things into perspective.

Friday January 5 2001			
	Service 1	Service 2	Service 3
Seattle			
Bldg 1	Up	Up	Down, up at 12 PM
Bldg 2	Down 8-10 PM	Down 8-10 PM	Down 8-10 PM
Bldg 3	Up (Slow)	Up	Up
Sydney			
Bldg 1	Up	Up	Down, up ?
Bldg 2	Up	Up (slow)	Up
. . .			

Figure 5.12 Service availability (reducing the element of surprise).

One often hears stories about somebody taking a car to the garage, complaining about sluggishness when accelerating or perhaps a peculiar knock under the hood. A mechanic hooks the engine to a computer and cannot find anything wrong. The mechanic argues that the car is in good shape—the computer said so. But the consumer insists that something is definitely awry. So the mechanic takes the car for a spin to get first-hand experience.

We have a semantic disparity problem here. If we apply the three approaches to the problem, we might have something like this: With the user-centric approach (taken to the extreme), the mechanic would accompany the car owner at all times, making sure the car is performing correctly. Clearly, that is impractical, almost laughable. With the happy-medium approach, the mechanic would take the car on occasional drives (say, once a month) to make sure everything is in order. With the techno-centric approach, the mechanic hooks the car to a computer.

With brand-new cars, the happy-medium approach and the techno-centric approach are taken together during periodic checkups. Suppose the former approach yields a problem that the latter approach does not corroborate (or vice versa). That sort of thing causes headaches, and finding a correct mapping between user-centric data and techno-centric data distinguishes excellent car mechanics from average ones. With that in mind, let us proceed to the component-to-service mapping problem.

5.3   The component-to-service mapping problem

The component-to-service mapping problem is one of finding a function or procedure that takes raw component parameters (device, traffic, system, and/or application parameters) as arguments and provides a value for an inferred, higher level service parameter. That is the challenge, of course, if we elect to take the techno-centric approach to the semantic disparity problem.

In general, we can look at the problem as follows:

$$f(P_1, P_2, \dots , P_n) = S$$

where the Ps are values of low-level component parameters, S is the inferred value of a higher-level service parameter, and f is the function that maps the Ps to S.

Assuming that we have defined S and the acceptable level for S, then the problem is reduced to selecting the Ps and defining f. We assume that f can include common arithmetic operators ($+$, $-$, $/$, $*$, $>$, $<$, min, max, and so on) and Boolean operators (and, or, not, if-then).

Chapter 1 introduced the component-to-service mapping problem with a common function for measuring service availability. We uncovered an obvious fault with the function but did not correct the fault. We will do that now and expand on the example.

Suppose seven components (say, three network devices, two systems, and two applications) combine to support a service. Assume we have monitoring agents in place for each of the seven components and the agents can measure the availability of their respective components.

It is tempting to say that the health of the service is acceptable if each component is available 98% percent of the time. However, the service could be unavailable 14% of the time, even though we would have to say that the metric that had been agreed on had been met. That is not right.

If A_n is the percentage of availability of component n over some given period of time, then the (faulty) function that describes this mapping is:

$$f(A_1, A_2, A_3, A_4, A_5, A_6, A_7) = \text{acceptable if } A_1 < 98\% \text{ and}$$
$$A_2 < 98\% \text{ and}$$
$$A_3 < 98\% \text{ and}$$
$$A_4 < 98\% \text{ and}$$
$$A_5 < 98\% \text{ and}$$
$$A_6 < 98\% \text{ and}$$
$$A_7 < 98\%$$
$$\text{else unacceptable}$$

One might be inclined to offer the following function in its place:

$$f(A_1, A_2, A_3, A_4, A_5, A_6, A_7) = \text{acceptable if } [700 - (A_1 + A_2 + A_3 + A_4 +$$
$$A_5 + A_6 + A_7)] < 2\%$$
$$\text{else unacceptable}$$

However, that function is faulty as well. If each component were available 98% of the time but exactly at the same time, then the 98% availability requirement will have been met. But the function above tells us it was not met. So this one is not right either.

A better function would look at the availability of each component as a time line, where gaps in the line show when the component was unavailable. If we imagine the seven lines superimposed on each other, where gaps override black space, the total availability is 100 minus the gaps (assuming normalization).

But now we can foresee another problem. Suppose a component (i.e., device, system, or application) was scheduled to be unavailable. We need to factor that into the function as well. We can do that by redefining A_n. Earlier we defined A_n as just the availability of element n. Now we define it as follows:

$$A_n = 100 - (UUA_n / SA_n)$$

UUA_n is a measure of unscheduled unavailability of component n (i.e., real downtime) and SA_n is a measure of scheduled availability of component n.

Now we are getting closer to what we want, albeit at the expense of introducing an extra burden on the monitoring agents. The agents have to know whether unavailability is planned or unplanned. The reader is invited to think further about other possible problems with our mapping function.

The remainder of this section describes a novel approach to the component-to-service mapping problem that is quite techno-centric. However, the scientists among us are urged to consider the following: Suppose we have a database transaction agent as described in Section 5.2, and we have taken the happy-medium approach to measuring service availability. We may ask whether we need to worry about the techno-centric approach in the first place.

Fuzzy logic

Current monitoring agents are very good at reporting values of parameters such as network load, packet collision rate, packet transmission rate, packet deferment rate, channel acquisition time, file transfer throughput, and application response time. Daemons can be attached to those parameters so that values that exceed a given threshold result in an alarm.

There are good graphics tools that can display that information in the form of bar graphs, X-Y plots, histograms, and scatter plots. However, except for clear faults, few capable experts can interpret those values and alarms in common sense terms and point to reasons for service degradations. Reasons for such degradations might include an overloaded network link, a router with an insufficient CPU, or an incorrectly adjusted timer for a transmit buffer. In addition, the task of detecting and correcting performance problems is becoming harder with the advent of increasingly large and heterogeneous networks.

One approach toward solving those problems is to simulate a service with a mathematical model. One can then predict the nature of services by running the model with simulated conditions. Unfortunately, most services do not lend themselves to mathematical modeling, either because they are too complex and dynamic to be modeled or because the computational expense of running the model is prohibitive.

A second approach is to simulate the expertise of a good network troubleshooter. The usual way to do that is to construct algorithms that translate streams of numeric readings of monitoring agents into meaningful symbols and to provide an inference mechanism over the symbols that captures the knowledge of the best experts in the field.

Most current implementations represent the requisite knowledge in an RBR framework (see Section 5.1). However, that sort of solution has shortcomings. This section describes an alternative implementation in which knowledge is expressed in a fuzzy logic framework.

A review of RBR systems is in order. Refer back to Figure 5.4. An RBR system consists of a working memory (WM), a knowledge base of rules, and a reasoning algorithm. The WM typically contains a representation of characteristics of the service, including topological and state information of components that support the service. The knowledge base contains rules that indicate the operations to perform when the service malfunctions.

If the service enters an undesirable state, the reasoning algorithm selects those rules that are applicable to the current situation. A rule can perform tests on enterprise components, query a database, provide directives to a configuration manager, or invoke another RBR system. With those results, the RBR system updates the WM by asserting, modifying, or retracting WM elements. The cycle continues until a desirable state in WM is achieved.

Several variations on the basic RBR paradigm exist. For example, the reasoning algorithm can be enhanced with a belief revision capability. The algorithm keeps a list of rules selected on each cycle and may backtrack to a previous cycle to select an alternative rule if progress is not being made toward a desirable state (assuming no operation has been performed that cannot be undone). In addition, the rule base can be functionally distributed, and a metacontrol strategy can be provided that selects the component RBR system that should be executed for specific kinds of tasks.

The usual procedure for constructing an RBR system is to (1) define a description language that represents the problem domain, (2) extract expertise from multiple domain experts or troubleshooting documents, and (3) represent the expertise in the RBR format.

The procedure can require several iterations of an interview-implement-test cycle to achieve a correct system. If the domain and the problems encountered remain relatively constant, a correct system needs little maintenance. However, if the system is used in unpredictable or rapidly changing domains, two problems inevitably occur: (1) the system suffers the problem of brittleness, and (2) the development process suffers the problem of knowledge acquisition bottleneck.

For example, this function describes a set of rules for issuing notices about the traffic load on a network link in an enterprise:

$$\text{notice} = \begin{cases} \text{alarm} & \text{if } load \leq 10\% \\ \text{alert} & \text{if } 10\% < load \leq 20\% \\ \text{ok} & \text{if } 20\% < load \leq 30\% \\ \text{alert} & \text{if } 30\% < load \leq 40\% \\ \text{alarm} & \text{if } load > 40\% \end{cases}$$

In that example, there is a WM element, $load$, that is monitored by a traffic monitor. The numeric value of $load$ is compared to the rules at prespecified time increments, and one rule fires to update the value of $notice$.

In some cases, the reading of a load's value along an interval of length 0.02 could make a big difference, whereas in other cases the reading of a value along an interval of length 9.98 makes no difference. For example, a value of $load = 9.99$ issues an alarm, and a value of 10.01 issues an alert, whereas the values 10.01 and 19.99 both issue an alert. Of course, that is so because the rule set describes a function that is discontinuous, as shown in Figure 5.13.

For issuing alerts and alarms, perhaps that is acceptable. However, the lack of continuity of a rule set becomes problematic for other useful variables. Suppose we are interested in the variable $reroute\%$, which tells us the percentage of traffic to reroute to maintain a notice of "ok." A possible implementation of that function is

$$\text{reroute}\% = \begin{cases} 5\% & \text{if } 30\% < load \leq 40\% \\ 15\% & \text{if } load > 40\% \\ 0\% & \text{otherwise} \end{cases}$$

That function is unsatisfactory; the primary reason is that the rules are "brittle." The antecedent (the "if" part) of a rule is either true or

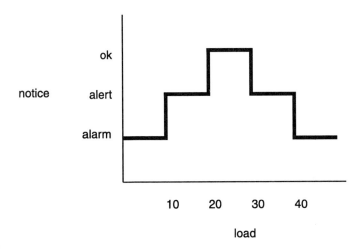

Figure 5.13 A graph of the rules for load notices.

false, and the output (*reroute%*) is always either 0%, 5%, or 15%. For example, 40% load returns 5% reroute, but 41% load returns 15% reroute. One approach to getting around the brittleness problem is to add more rules. That approach, however, is likely to result in a proliferation of rules and thus introduce the knowledge acquisition bottleneck problem.

Those problems are grounded in what we call the "lean semantics" of the RBR approach. We would like to describe the load of a network in terms like "heavy," "very heavy," "slightly heavy," and so on. We would like to look at a load measurement of say 29% and say that it is "not heavy but not ok either" or simply "slightly heavy."

The fuzzy logic framework described in this section allows us to translate numeric performance variables in such common sense terms. It enjoys a rigorous mathematical underpinning and affords us a richer semantics than the RBR framework.

On first glance, the fuzzy logic paradigm seems to provide just the apparatus we need to solve the semantic disparity problem and the component-to-service mapping problem in one fell swoop. This is a current research area, so only time will tell whether the approach bears out in practice.

With the fuzzy logic approach, we construe network parameters as reported by monitors (e.g., load, collision rate, response time) as linguistic variables and provide membership functions that translate the parameters' numeric values into degrees of membership in a fuzzy set. A linguistic variable is a variable whose values range over linguistic terms rather than numeric terms. The variable *load* might range over the common sense terms "light," "ok," and "heavy."

A simple comparative illustration is as follows. Let the variable *load* have a universe U over the interval [0,100%]. Figure 5.14(a) shows the concept "heavy" in crisp logic (i.e., nonfuzzy, two-valued logic). With the fuzzy logic framework, we could define a fuzzy set over U that describes the common sense term "heavy" with the following function:

$$heavy = \begin{cases} 0.0 & \text{for } 0 < y \text{ in U} \leq 25 \\ \left\{1 + \left[(y - 25)/5\right]^{-2}\right\}^{-1} & \text{otherwise} \end{cases}$$

Figure 5.14(b) illustrates the fuzzy concept "heavy." A numeric value of *load* less than 25 would have 0.0 grade of membership in the concept "heavy," a value of 30 would have 0.5 grade of membership, and a value of 40 would have a 0.9 grade of membership.

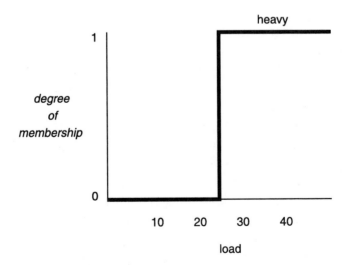

Figure 5.14 The concept "heavy" in (a) crisp logic and (b) fuzzy logic.

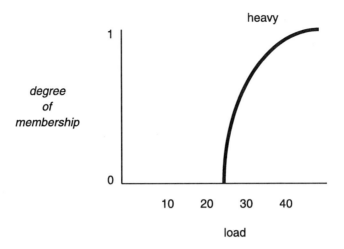

Figure 5.14   continued.

In similar fashion, we define fuzzy sets for the concepts "ok" and "light." Figure 5.15 shows what those functions might look like. Note that a value of 25 would have a 100% grade of membership in the concept "ok" but would have 0.0 in "light" and "heavy." Also note that now a value of 30 would participate to degree 0.5 in "heavy" and about 0.8 in the concept "ok."

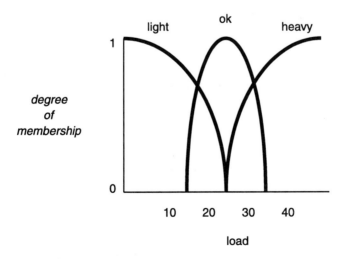

Figure 5.15   The concepts "light," "ok," and "heavy."

Figure 5.16 shows the general engineering methodology for building and fine-tuning a fuzzy logic system. First, we define a grammar representing (1) input variables from monitoring agents (e.g., load, packet transmission rate, channel acquisition time, availability, and response time) and (2) output variables (notices, service health, network load adjustment, and transmit buffer timer adjustment). Next, we define membership functions for each concept. Then, we allow experts to define fuzzy rules that connect input variables and output variables, while the system builder selects a fuzzy inference strategy. The "defuzzification" uses the same member functions to translate common sense terms back into numeric terms.

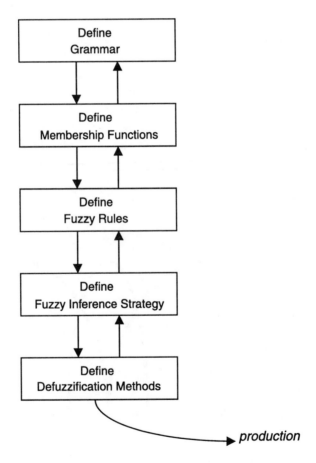

Figure 5.16 The engineering framework for building fuzzy logic systems.

Examples of fuzzy rules follow.

- If load is heavy and file_transfer_throughput is slow then service_health is weak and bandwidth_adjustment is small increase.

- If load is not heavy and packet_collision_rate is high then transmit_buffer_timer_adjustment is small increase.

- If load is very heavy then notice is strong alert and reroute% is medium decrease.

- If load is medium and rate_of_load_change is high increase then notice is alert and reroute% is small decrease.

Figure 5.17 shows the operation of a fuzzy logic system for service management once it has been built. The horizontal line in the figure

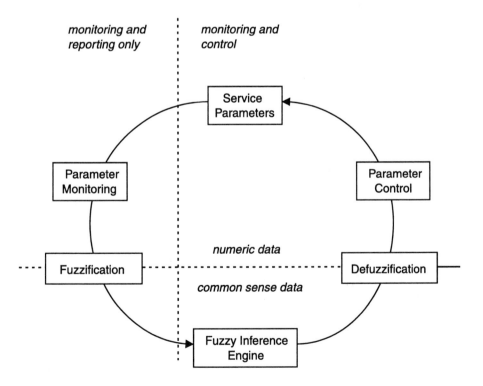

Figure 5.17 The operation of a fuzzy logic system for service management.

shows the separation of numeric data and common sense data. The vertical dotted line suggests that we can build a fuzzy system that performs monitoring and reporting only, as opposed to one that also performs service control.

Consider the fuzzy inference engine. All antecedents of fuzzy rules that participate in the "truth" of the input data will fire and thus contribute to the overall solution. Further, an antecedent does not have to be an exact match with the input data.

The output variables of a rule are tweaked relative to the degree of match between the antecedents of the rule and the (fuzzified) input of parameter monitors. The most common fuzzy inference mechanism is called the *compositional rule of inference*. We will not describe the underlying calculations here. The Further Studies section at the end of this chapter lists literature that contains good discussions of standard formulas for generic fuzzy concepts and standard inference mechanisms. In addition, the literature contains case studies of successful applications of fuzzy logic to domains other than service management.

Where are we?

At this point, let us take stock of our line of reasoning so far. First, we described the semantic disparity problem, namely, that user views and technical views of services are different, and it is hard to reconcile them. We discussed three approaches to the problem: the user-centric approach, the happy-medium approach, and the techno-centric approach.

Paying heed to good SE practices, we argued that the user-centric or happy-medium approaches should be preferred and investigated to the fullest. For example, service availability can be monitored and measured by a representative set of transactions by a special purpose agent.

Having made that point, we nonetheless proceeded to discuss the techno-centric approach in more detail. First, we looked at standard procedures for mapping component parameters to service parameters. Then we described an innovative approach to the mapping problem that would appear to be especially suited for it: fuzzy logic. The fuzzy logic approach shows promise, but it is pure research at this stage; while it is interesting and enticing, one should always favor the user-centric approach.

5.4 The agent selection problem

Refer to the essential SLM methodology in Figure 3.1. Particularly, note part (1) of step 7 in the methodology:

> Step 7. The supplier takes first steps toward (1) *identifying agents to monitor and control components*, (2) designing agent integration, and (3) experimenting with nonproduction prototypes.

Assuming that we have defined services in which a consumer is interested and have identified the components and component parameters on which the services depend, then we have the problem of finding suitable agents that can actually monitor the parameters.

We have touched on that problem at several places in preceding parts of the book and discussed some approaches to alleviate it. This section brings it all together.

First, let us review the general categories of monitoring agents.

▶ *Device agents* (also known as element managers) have a focused view of the connection nodes in the network infrastructure, for example, bridges, hubs, routers, and switches. Service parameters typically include port-level statistics. Good examples in the industry include Cisco's CiscoWorks, Bay Network's Optivity, and Cabletron SPEL. A hardware vendor that produces a network device is obliged to build a management agent for the device.

▶ *Traffic agents* have a focused view of the traffic that flows over transmission media in the network infrastructure. Examples of service parameters include bytes over source-destination pairs and protocol categories thereof. Examples in the industry are HP Net-Metrix, NetScout's RMON, and NDG Software's Programmable RMON II+.

▶ *System agents* have a focused view of the systems that live in the enterprise. Typically, these agents reside on the system, read the system log files, and perform system queries to gather statistics. Good examples are Metrix WinWatch, BMC Patrol, Tivoli TME, and

CA TNG. Service parameters include CPU usage, disk partition capacities, and login records.

▶ *Application agents* have a focused view of business applications that live in the enterprise. These agents also reside on the system that hosts the application. Good examples are Optimal Application Expert, Metrix WinWatch, Platinum ServerVision, and BMC Patrol. Some applications offer agents that provide indices into their own performance levels. Service parameters include thread distribution, CPU usage per application, file/disk capacity per application, response time, number of sessions, and average session length.

▶ *Special-purpose agents* monitor parameters that are not covered by any of the above. A good example is an agent whose purpose is to issue a synthetic query from point A to point B and (optionally) back to point A to measure availability and response time of an application. Note that the synthetic query is representative of authentic application queries. An example is an e-mail agent that monitors the response time and jitter of e-mails from one user domain to another. Good examples in the industry are Optimal's Application Expert, HP's PerfView/MeasureWare, and Jyra's Jyra.

▶ *Enterprise agents* have a wide-angle view of the enterprise infrastructure, including connection nodes, systems, and applications that live in the enterprise. These agents are also cognizant of relations among the components at various levels of abstraction and are able to reason about events that issue from multiple enterprise components. This is called event correlation or alarm rollup. Service parameters that are accessible by enterprise agents are numerous, including router and hub statistics, ATM services, frame relay services, and link bandwidth usage. Examples in the industry are Cabletron Spectrum, HP OpenView, and IBM NetView.

We may consider those agents as real-time agents. They manage the enterprise on a daily basis. Chapter 4 introduced off-line agents that look at management data from a historical perspective for purposes of maturing and supporting services over time. Compare Figures 4.7 and 4.10. The case study at the end of this chapter provides a good example of an off-line agent that performs data mining on historical data to find ways to improve service delivery.

The sequential steps of the SLM methodology in Figure 3.1 assumed that we are starting with a clean slate regarding SLM development and deployment. However, it is hardly ever the case (nay, *never* the case) that a company that embarks on SLM will not have some management agents already in place. Chapter 3 discussed several variations on the essential SLM methodology that bear on the agent selection problem, and a short review is in order.

The backtracking strategy involves moving back from step 7 to prior steps if we insist on using the existing management system and the system does not include appropriate agents for managing the services of interest. That approach calls for a reidentification of services of interest.

The version N space strategy can be considered a special case of backtracking. If a number of services are defined but the existing management system does not include appropriate agents for managing all the services, then we can pick out those services the system can accommodate and call that version 1 space.

The starting at the bottom strategy is to begin with the existing management system and work backward to identify the services that can be monitored and controlled.

The starting at the top and the bottom at the same time strategy starts with the identification of services and the identification of existing management procedures. We then work our way toward the middle of the methodology, hoping they connect. Usually, however, they do not.

Chapter 4 discussed how the management styles of some businesses, for whatever reason, are network-, systems-, applications-, or traffic-centric. A further observation, not discussed in Chapter 4, is that these styles are not necessarily exclusive.

For example, enterprise management systems such as Spectrum and OpenView use autodiscovery mechanisms to formulate a topology of the enterprise at various levels of abstraction. But it is possible to construct a simple form of enterprise topology with traffic monitoring agents by reading source and destination IP addresses in packets that flow over the enterprise. However, it is very unlikely that the two topologies will be identical. First, traffic agents generally do not include logical components of enterprises. Second, traffic agents do not see intermediate components such as switches in a switch fabric.

For a second example, consider that we can measure LAN traffic with a traffic agent. But we also can do that by monitoring the ports of a router with a device monitoring agent or an enterprise agent.

For a third example, consider that enterprise agents such as Spectrum and OpenView exhibit a simple form of computer systems management. They can measure the availability of the systems, collect SNMP statistics (assuming that an SNMP agent resides on the system), and perform event correlation over systems and network devices. On the other hand, the management capabilities of system agents such as BMC Patrol and Tivoli TME are a rather large superset of this, minus the ability to perform event correlation over systems and network devices.

As a final example, consider the different ways to measure application response time. VitalSign's Vital Agent resides on the desktop and measures packets in and packets out to calculate response time. Envive's StopWatch resides next to a server, mirroring the server's ports and uses Sniffer technology to track data in and out of the system for a given application. Jyra's Jyra agent periodically runs a set of transactions and records/trends response times. And Candle's ETEWatch resides on the desktop and monitors Windows events to measure response time.

How do such observations bear on the agent selection problem? The answer is obvious. Crudely put, there are more ways than one to skin a cat. Keep that in mind as we work through step 7 of the SLM methodology.

5.5 The integration problem

The integration problem focuses on part (2) of step 7 in the SLM methodology:

> Step 7. The supplier takes first steps toward (1) identifying agents to monitor and control components, (2) *designing agent integration*, and (3) experimenting with nonproduction prototypes.

The integration problem is common in both the telecommunications industry and the enterprise management industry. Typically, there are several management systems in place that operate independently of each other, although a systems analyst or a consumer will see clear benefits if the systems are integrated in some fashion.

Integration projects are motivated by the fact that no one vendor can provide all the solutions for enterprise management. Vendors usually are specialists in one area, for example, network, systems, application, or traffic management. Therefore, it often is necessary to integrate multiple systems into a unified whole to have a comprehensive enterprise management system.

We have touched on the integration problem at several places in this book. Recall that in our discussion of SE practices in Chapter 3 we made a distinction between an analysis model and a design model. We argued that an analysis model should be constructed without regard to environmental constraints or available tools; it describes an ideal system.

When we transform the analysis model into a design model, we face reality and find that we may not be able to build the ideal system. Nonetheless, we carry over what we can and have good pointers for further research and study.

The integration problem is a design problem and thus posterior to analysis. In Section 5.1, on distributed event correlation, we followed that practice properly. Consider the discussion of the distributed, intercommunicative management system in Figure 5.11. We assumed that all individual management systems could, in theory, communicate with each other. Then we proceeded to outline scenarios in which communications among agents would be beneficial. With those scenarios in hand, we now have the job of finding communication mechanisms to make it happen.

The integration problem is widely recognized in the industry. Several standards bodies and industry consortia are working on common protocols and languages by which management agents can communicate. For example, from an implementation perspective, the OMG has selected CORBA as an implementation mechanism between diverse objects in a management system.

From a conceptual perspective, a U.S. consortium is developing the common information model (CIM), which describes a common language to be shared by objects in a management system. Often it turns out that a consortium developing a conceptual language such as CIM adopts an implementation mechanism such as CORBA.

Meanwhile, vendors who develop management agents have developed public interfaces though which their agents can receive and request information from other agents. At the same time, they are looking at the work in standards bodies and industry consortia scattered over the globe. Vendors do not want to be left behind if work on a standard really becomes a standard.

From this discussion, we can glean two approaches to the integration problem: the standardized approach and the vendor-specific, public interface approach.

Fortunately, the two approaches are not mutually exclusive. A vendor's product may have a public interface that expedites integration with other management agents nicely. Now, if a standard becomes commonplace and accepted in the industry, the vendor may have to provide a wrapper around its public interface so that it speaks the same language as everybody else in the community.

Consider a simple example in which an analysis model calls for the passing of alarm information from a peer management agent to the Spectrum enterprise agent. When Spectrum was first built in the early 1990s, it was equipped with a C++ application programming interface (API). The C++ API was in turn used to develop a command line interface (CLI) in Spectrum.

The CLI is a useful tool for implementing an integration based on an analysis model. However, if the CLI mechanism does not provide the necessary functionality, we can revert to the C++ API. Now, in the late 1990s, Spectrum is equipped with a CORBA interface. Thus, there are three mechanisms by which peer agents can communicate with Spectrum.

In sum, the integration problem is not an insurmountable one. Most management agents provide public interfaces in one form or another. The prior problem is more important—understanding the content of the communication.

5.6 The scaling problem

The scaling problem with respect to SLM is this: Consider *end-to-end* SLM, in which we try to cover every possible component that could affect a particular service. Then consider *selective* SLM, in which we cover only a select few of all possible components that could affect a particular service. The problem is one of correctly selecting the few components while still being assured that we are measuring the service.

The distinction between end-to-end and selective SLM is reminiscent of a common distinction in AI between precise, complete solutions and "satisficing" solutions. A satisficing solution to a problem is one that is considerably less expensive than, but not far from, a precise and complete solution.

The scaling problem is related to two problems already discussed: the semantic disparity problem and the component-to-service mapping problem. We distinguished among three approaches to those two problems: the user-centric approach, the happy-medium approach, and the techno-centric approach.

For example, the scaling problem shows itself when we take the techno-centric approach to availability measurement. One way to alleviate the scaling problem is to bypass the techno-centric approach altogether and instead to find some way to directly measure a service from the user's point of view. Earlier we discussed a representative transaction agent that measures availability and response time of users' applications.

The case study at the end of this chapter considers a rather advanced method toward alleviating the problem. We use data mining algorithms to discover the critical components on which a service depends. For example, if response time is a measure of a service, we can compare the measurements of response time to measurements of all other component behaviors. In that way, we may find a close correlation between response time and some critical component or set of components in the enterprise.

5.7 The representation problem

To understand the representation problem, consider a simple analogy, a football team. It is straightforward to categorize the players who make up the football team into offensive players, defensive players, running backs, line backers, and so on. It also is straightforward to look at the health of each individual player and to aggregate those individual healths into a single measure of the team's overall health.

In doing that, however, we do not have a representation of a football play. Obviously, the excellent health of a football team does not guarantee that a football play also will be excellent. Something is missing from the representation, namely, the process of playing football and the evaluation thereof.

We can apply those ideas to our understanding of a service. It is one thing to represent a service in terms of the components that support it, but it is quite another thing to represent a service as a process. Process representation is much harder, albeit quite useful if one can do it.

Figure 5.18 shows a hand-drawn topology of a complex enterprise for Company X, with various kinds of networking technologies, satellite communications, file servers, and workstations.

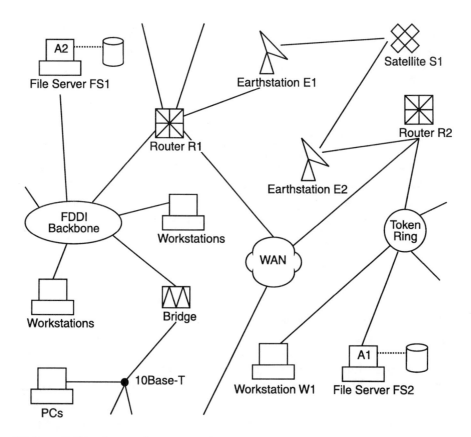

Figure 5.18 A complex enterprise.

Suppose Company X has a policy to update file server FS1 (in the top left portion of the figure) every Wednesday with the prior week's accounts received, which have been logged in file server FS2 and a home-grown database on workstation W1 (bottom right). This is an example of a service, which we will name "weekly account aggregation."

A component representation of the service would group the components on which the service depends in a single, logical view. The health of the service, then, is a function of the health of the underlying components. Looking at Figure 5.18, we easily can pick out the components on which the service depends.

A process representation of the service would model the movement of data over the enterprise. For example, suppose the retrieval of data D1 from W1 initiates the process, and D1 is sent to application A1 on FS2. A1's responsibility is to combine D1 with data D2 retrieved from FS2 (say that D3 = D1 + D2), format it, and ship D3 to FS1 for permanent storage. Application A2 on FS1 performs a security check on D3 to make sure it has not been corrupted or tampered with en route and then deposits it in the database on FS1.

Figure 5.19 shows a possible representation of the service using the STG paradigm (refer to Section 5.1). Note that the STG depends on application agents that forward application events to the representation.

Clearly, there are other ways to represent the weekly account aggregation service. We can review the other reasoning paradigms discussed in Section 5.1 to try to determine whether there is a better way to do it. For example, the service may be cast as a set of business rules, and as such the process could be represented in an RBR system.

5.8 The complexity problem

The complexity problem is perhaps the hardest challenge of all. Suppose we have a service in a business that is crucial. Chapter 1 cited a distributed aircraft scheduling service as an example. Another good example is a telemedicine service that involves on-line diagnosis supported by electronic scan images, pathology test results, patient records, and communication among medical staff.

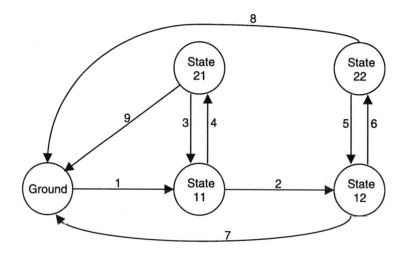

Ground: No Activity

State 11:   A1 in progress
State 12:   A2 in progress
State 21:   A1 failed
State 22:   A2 failed

Arc 1:   A1 start event
Arc 2:   A2 start event
Arc 3:   A1 restart event
Arc 4:   A1 failure event
Arc 5:   A2 restart event
Arc 6:   A2 failure event
Arc 7:   A2 completion event
Arc 8:   A2 user abort event
Arc 9:   A1 user abort event

Figure 5.19   An STG for weekly account aggregation.

Clearly, we want those kinds of services to be running at peak performance at all times. However, we know that in full these applications depend on various kinds of network devices, computer systems, traffic links, and distributed client/server applications. They are good candidates for end-to-end management.

We can partly guarantee good performance by following good SLM methodological practices. In other words, we can quantify service performance for a new distributed application in a pilot test before putting the application into production. Consider the following typical example.

Figure 5.20 shows a high-level view of an enterprise that consists of sites A, B, C, and D and a WAN. Suppose we want to install a distributed application over the enterprise, where a server farm is to be located at site A and its clients are to be distributed over all sites.

Before installing the servers and clients, we measure the traffic utilization between each site to get a baseline. Then, we install the servers and clients. We simulate the load that is expected to be caused by the application between sites by issuing an appropriate set of generic database transactions. During the simulation, we take a second measurement, in which we measure again the traffic utilization between sites and also the response time of the representative set of transactions.

Note that there are two things we want to understand with this pilot test: (1) whether the existing enterprise can accommodate the extra load caused by the new database application and (2) whether the response time of the application is acceptable. If not, there are several kinds of solutions we can try, depending on the existing enterprise configuration (e.g., replacing or upgrading some sites with alternative networking technologies).

This example should cause several kinds of worries. First, the approach is rather traffic-centric. Second, there is an implicit assumption that the application does not affect other applications that already exist in the enterprise. Third, there is the question of whether the simulation adequately simulates the real production system. And finally, no matter how hard we try to test new applications before putting them into production, there always seem to be glitches that were unanticipated.

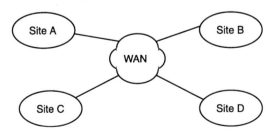

Figure 5.20 An enterprise with four sites connected to a WAN.

That is just what we mean by the complexity problem. Assuming we have a distributed service in place and a way to measure its performance, our question is simply this: What conditions affect its performance and to what extent? Because the service depends on network devices, traffic loads, computer systems, and applications, that is a hard question to answer.

If we can uncover the conditions that lead to good and bad performance of a service, we are halfway toward ensuring that the service is reliable and trustworthy. The case study in Section 5.9 describes a method for doing that: data mining.

5.9 Case study:
KLM Airlines

This section describes an experiment at KLM Airlines (Netherlands) in which data mining techniques are applied to historical data collected by monitoring agents. A service is identified and measured based on response time. Large amounts of performance data, including network, traffic, system, and application performance data, are collected by a set of monitoring agents and stored in a data repository. Then data mining algorithms are executed on the data to learn causal relations between other parameters in the enterprise and the response time parameter.

The case study was pioneered by researchers and engineers at Syllogic B.V. (Netherlands). Because a fair amount of research and experimentation preceded the case study, we will spend some time describing the prior work before we enter into the case study. The agenda for this section is as follows:

▶ Motivation for data mining;

▶ Concepts, terms, and algorithms in data mining;

▶ Three preliminary lab experiments;

▶ Finale: the case study at KLM Airlines;

▶ Discussion: data mining in enterprise management.

Motivation for data mining

As enterprises become larger and increasingly heterogeneous, it becomes difficult to manage the complex interactions between different components such as network devices, databases, computer systems, and applications. Each component can affect the operation of the other components and ultimately affect the performance of the services and BPs they support.

Some events that cause the degradation of service performance or a BP in an enterprise are simple and can be resolved easily or prevented by a human troubleshooter. However, with the increasing complexity of modern enterprises, it is becoming harder to pinpoint the exact cause of some service degradations.

In many cases, bad performance is not caused by a single, sudden event, but by a subtle interplay between several performance parameters that causes a gradual change in service performance. Thus, service management becomes a complex diagnostic problem instead of a set of simple reactive operations.

Although it is hard for the human mind to get clear insight into the dynamics and mutual influences among the components that support a service, it is relatively easy to monitor the different components with current monitoring agents and to collect raw performance data in a data repository. Now, given the availability of such historical data, the application of data mining techniques to extrapolate usable management information is an obvious next step.

The goal of data mining in enterprise management is to transform large amounts of raw data into information or knowledge that can be comprehended and used by enterprise administrators. For example, the knowledge may take the form of discovering cause-and-effect relationships among components in a system or being able to discover particular component parameters that distinguish a healthy service from an unhealthy service.

Concepts, terms, and algorithms in data mining

The first requirement for a data mining application is to collect and store data that describes the state of a system at regular intervals. The data can include configuration data, events and alarms, and, more important, performance data.

The data collected by a set of agents are organized into a time-ordered set of *parameter vectors.* The monitoring agents combine to produce parameter vectors that reflect the state of the system at particular time increments or over an interval of two measurements.

Consider a service level parameter such as *response time.* If we collect a month's worth of parameter vectors at 10-minute intervals, where *response time* has been designated as the service level target, we can apply data mining algorithms to discover how other parameters influence the behavior of the response time parameter. Clearly, that would be useful knowledge.

Let us distinguish between two representations of such knowledge: *propositional* and *quantified* representations. The former is associated with propositional logic, the latter with quantifier logic (also known as predicate logic or first-order logic).

In propositional logic, the unit of what we can say is a whole sentence, although we may use the usual Boolean operators to create complex sentences. In quantifier logic, our units of description are objects and predicates, and we are allowed to make *universal* and *existential* statements that range over sets of objects.

For example, consider the complex sentence "r4 is an AIX server and r4 resides in domain 1." In propositional logic, that fact can be represented by the statement *P and Q*, where:

$$P = \text{"r4 is an AIX server"}$$
$$Q = \text{"r4 resides in domain 1"}$$

In quantifier logic, the same statement can be expressed as *Kab and Rac* (by convention we put a predicate in front of the objects to which it applies), where:

$$K = \text{"is a kind of"}$$
$$R = \text{"resides in"}$$
$$a = \text{"r4"}$$
$$b = \text{"AIX server"}$$
$$c = \text{"domain 1"}$$

Further, in quantifier logic we can express concepts such as "All AIX servers reside in domain 1," and "At least one AIX server resides in domain 1." Those two statements express a universally quantified statement and an existentially quantified statement, respectively, and they can be stated in quantifier logic as follows:

> For all x: if Kxb then Rxc
> There exists an x such that: Kxb and Rxc

The preceding discussion illustrates the differences in expressive power of propositional statements and quantified statements. Now, some data mining algorithms discover propositional knowledge, while others discover more general quantified knowledge. The latter is useful because it is more general and has more predictive power. The trade-off, however, is that the latter requires background domain knowledge and is somewhat computationally expensive.

Decision tree algorithms produce propositional knowledge in the form of a decision tree. Figure 5.21 shows a generic decision tree in which each node in the tree is a proposition. The algorithm takes a large table of data as input, in which a service parameter (SP) is marked as the target parameter. The algorithm produces a decision tree that shows the major influences on SP. By starting at SP at the root of the tree, we can examine important dependencies by proceeding toward the leaves of the tree. Popular algorithms of this kind are ID3 and its derivative C4.5.

Top N algorithms produce propositional knowledge as a simple list that shows the top N parameters that are the major influences on the target service parameter, in decreasing order of influence. Unlike decision trees, top N algorithms do not uncover dependencies on multiple influential parameters. *Rule induction algorithms* produce propositional knowledge in the form of rules that show the dependencies between a target parameter and multiple influential parameters. Consider this example of such a rule:

> *if* CPU idle time on r4 > 63%
> *then* response time > 2 seconds

The statement is indeed useful; however, we are inclined to ask further questions: Are there other machines for which the rule also holds?

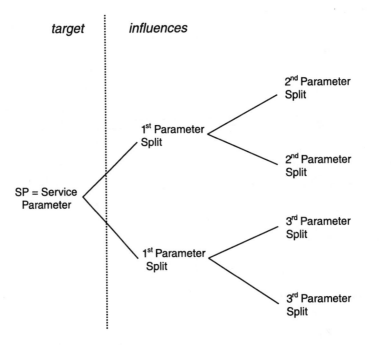

Figure 5.21   The structure of a table and a derived decision tree.

Are there classes of machines for which the rule holds? Are there instances of such classes in my enterprise? The answers to questions like these will be quantified statements instead of propositional statements. For example:

> For all x: if x is and AIX server and
> CPU idle time on x > 63%
> then response time > 2 seconds

Inductive logic programming (ILP) algorithms produce quantified statements by appealing to domain knowledge in addition to knowledge collected in a performance table. Such domain knowledge includes the relationships known to hold in the domain of the enterprise, for example, componentwise relations and hierarchical decompositions of components into subcomponents. For example:

> r4 is a kind of AIX server
> All AIX servers are kinds of UNIX servers
> CPU idle time is a parameter of a UNIX server

Domain knowledge is used by ILP to infer more general knowledge. The statements of the knowledge discovered by ILP algorithms can include both propositional knowledge and quantified knowledge. For example:

(propositional) If CPU idle time on r4 is ...
(quantified)   If x is an AIX server and CPU idle time on x is ...

Although statements of the first type are useful, quantified statements of the second type come closer to what we mean by knowledge. Also, they are more general and thus more useful in diagnosing related enterprise problems. Given those distinctions, we can characterize three data mining tools that were used in the preliminary lab experiments and the case study at KLM.

▶ The *Adaptive System Management* (ASM) tool, developed at Syllogic B.V., contains the three propositional algorithms described earlier (decision tree, top N, and rule induction).

▶ *Progol*, developed at Oxford University Computing Laboratory, is an ILP-type system that uses a rule-induction algorithm.

▶ *Tilde*, developed at the University of Leuven (Belgium), is an ILP-type system that uses the decision tree algorithm.

Three preliminary lab experiments

This section describes three preliminary lab experiments that led up to the real-world case study at KLM Airlines.

The goal of the first experiment, *File Access Influence*, is to discover the parameters that influence the access time to files in a file system distributed over several hosts. To do that, a data set is produced by periodically executing several UNIX commands over an interval of 30 minutes and measuring the response time. The result is a data set of 5,052 parameter vectors, where each vector consists of the following:

Weekday:	Sunday, ... , Saturday
Directory:	name of directory
Local:	location of directory
Filehost:	name of host
Dirsize:	total size in kB
CPU_usage:	load on local host
Response time:	seconds

The parameter response time (RT) is the target service parameter. The service level is defined to be acceptable if RT is less than 2 seconds; otherwise, it is unacceptable.

Figure 5.22 illustrates a decision tree (using ASM) that takes RT and the Directory parameter as input. Each node in the tree is labeled with (1) the parameter/value relation, (2) the total number of vectors with which the parameter/value relation is compared, and (3) the percentage of vectors for which the parameter/value relation is true. For example, one can see that RT 2 is true in 36.9% of the 5,052 vectors (i.e., 842 vectors). Further, of those 842 vectors one can see that Directory = /home/test/src/* and Directory = /usr/bin always result in RT > 2.

Note that the tree in Figure 5.22 excludes information about the other parameters. When the algorithm takes all parameters as input, Figure 5.23 is obtained. That tree tells us that the influences on RT > 2 are Local = False (i.e., the directory is a remote directory) and Dirsize > 2652 kB. These conditions hold in 100% of the cases in which RT > 2.

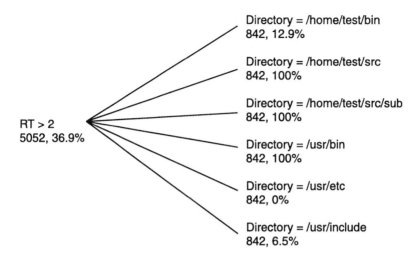

Figure 5.22 A two-parameter decision tree (file access influence).

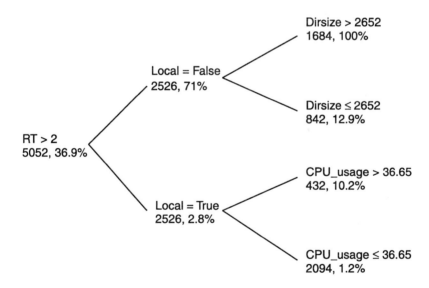

Figure 5.23 A multiparameter decision tree (file access influence).

The second tree clearly is more informative, albeit at the expense of extra work on the decision-tree induction algorithm. Consideration of only the Directory parameter in Figure 5.22 overshadows the combined influences of parameters in the system as a whole. The experiment tells us that the more parameters we consider, the more likely we are to find true influences on the target parameter of interest.

The goal of the second experiment, *Misbehaving Subnet*, is to discover the parameters that influence an overloaded subnet. The testbed for the experiment was a configuration of 16 subnets connected by a single router, where subnet 5 was known to be problematic.

Over a period of 18 weeks, a total of 16,849 vectors were collected for each subnet, where each vector consists of:

Day: the day on which the measurement was taken
Time: the time, in 10-minute intervals, at which the measure-
         ment was taken
LoadN: percentage of bandwidth utilization on subnet N
PktsN: packet throughput on subnet N
CollN: number of collisions on subnet N

Thus each vector contains 50 parameters:

Day, Time, Load1, Pkts1, Coll1, ..., Load16, Pkts16, Coll16

The data was analyzed in an attempt to pinpoint a problem causing complaints about occasional poor performance on subnet 5. At times, the users on subnet 5 would complain about sluggishness of their applications. The problem would disappear after about an hour.

A preliminary investigation showed that users started complaining whenever the load on subnet 5 started creeping above 35%, so a service parameter Load5 is defined such that Load5 less than 35% is acceptable; otherwise, it is unacceptable. Before the analysis was started, the parameters Pkts5 and Collision5 were removed from the data set because they are tied directly to Load5.

The decision tree (using ASM) in Figure 5.24 shows the dependencies between the target parameter Load5 > 35% and the remaining parameters. The poor performance of subnet 5 seems to be related to the load on subnet 9.

The dependency between subnets 5 and 9 suggests that there is a client/server communication going on between the two subnets. If such a communication is the main cause for the high load on subnet 5, there should be a roughly linear relation between the loads on the two subnets.

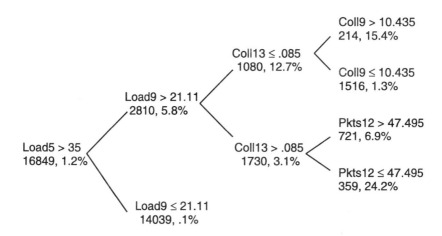

Figure 5.24 A multiparameter decision tree (misbehaving subnet).

To test that hypothesis, the ASM top N algorithm was executed on all pairs (Load5, LoadN), resulting in the following top three correlations:

$$\text{Load5 Load9 } 0.788$$
$$\text{Load5 Load1 } 0.131$$
$$\text{Load5 Load4 } 0.105$$

That test corroborates the hypothesis that subnet 9 strongly influences the misbehaving subnet 5. The experiment informs us that employing multiple data mining algorithms, in conjunction with knowledge of a network specialist who can interpret the results, can help us pinpoint service degradation problems.

The goal of the third experiment, *Behavior Modeling*, is to apply data mining algorithms to historical data in an attempt to construct a state transition graph (STG) that models enterprise behavior. That is quite an ambitious goal. However, if it is possible, the learned STG could be used for proactive and reactive service management and automated monitoring and control.

The data set for the experiment is the same as the one used in the Misbehaving Subnet problem. We will not go into the details of the experiment, because they are rather involved and open to interpretation. The motivated reader is referred to a research paper cited in the Further Studies section.

Finale: The case study at KLM Airlines

At KLM Airlines, the focus is on a particular service named "spare part tracking and tracing for aircraft," or SPT for short. The SPT service depends on several IBM AIX servers, an Oracle database, and Windows PC clients situated in Amsterdam, Singapore, and New York.

In total, monitoring agents were put into place that collected values of 250 parameters at regular intervals. Examples of some of the parameter types are *cpu load*, *free memory*, *database reads*, and *nfs activity*. The agents performed a read every 15 minutes and stored the values in a data warehouse. The SPT service was monitored for 2 months, resulting in a table of 3,749 vectors, where each vector consists of 250 parameters.

SPT performance was measured by simulating a generic transaction on the Oracle database and recording the response time of the transaction. The performance measure was declared as the pivotal measure in an SLA between the IT department and the users of the SPT. The determinator of good and bad performance of the SPT is governed by the test RT > 3 seconds. That means an SPT user should never have to wait more than 3 seconds before receiving the results of the transaction.

First, we consider the results of the propositional algorithms in ASM. The tree in Figure 5.25 shows the results of the decision tree algorithm. The most influential parameter is "Server 11 paging space." The tree tells us that a high value of that parameter is the main influence on RT > 3.

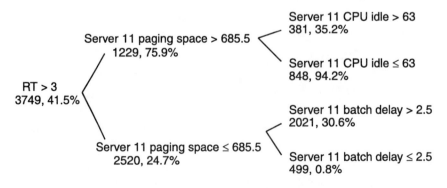

Figure 5.25   A multiparameter decision tree (SPT).

Increasing the amount of physical memory or limiting the number of applications that run on Server 11 can reduce the amount of used paging space. The next split on "Server 11 CPU idle" gives additional evidence for the fact that Server 11 needs to be upgraded or restricted to fewer applications.

Note the path from "RT > 3" to "Server 11 paging space ≤ 685.5" in 24.7% of the cases. The next parameter in the path, "Server 11 batch delay," measures the delay on scheduled jobs experienced by Server 11. Mainframe requests are sent (in batch) to a database that is accessed by Server 11 and then processed by Server 11. The split on "batch delay" suggests that if Server 11 is more than 2.5 minutes late in processing the batch file, SPT performance drops.

A seasoned troubleshooter who tries to make sense of that information might reason as follows: First, the network could be down, causing the mainframe to fail when it tries to send requests to the database, while at the same time causing Server 11 to time out because the query is performed over the network. Second, Server 11 could be wasting CPU cycles trying to retrieve a file that is not yet there, because the mainframe application has not yet put it there. In any case, the split on "Server 11 batch delay" indicates that the way Server 11 works with the mainframe should be improved.

Now compare the results of ASM's Top N algorithm on the same data, showing the top parameters that influence RT > 3:

Server 11 paging space > 685.5 MB
Client 6 ping time > 258.5 ms
Server 5 CPU idle < 74.5 %

The parameter "Server 11 paging space" corroborates the results of the decision-tree algorithm.

The parameter "client 6 ping time" is the ping time to a foreign router. It tells us that if the ping time exceeds 258.5 ms, then RT > 3 is likely to be true. A system manager at KLM reasoned that that fact could be related to foreign users who load complete tables from the database to their client. Because a table can be very big, and the network connections to foreign countries have a narrow bandwidth, both ping time and SPT behavior could be affected.

The parameter "Server 5 CPU idle < 74.5" is an influence on RT > 3, but to a lesser extent than the first two parameters. More important, observe that "Server 11 CPU idle ≤ 63" in Figure 5.25 is also a strong partial influence on RT > 3.

Now we compare the results of ILP algorithms used in Progol and Tilde. Recall that ILP-type systems appeal to a domain model to discover quantified knowledge.

First, note that ILP algorithms are CPU intensive. To compensate for that, the values in the original performance table were transformed into a table of binary values using well-known techniques. The loss of information in this preprocessing step should be considered a simplifying assumption.

Progol produced the following two rules from the transformed data set:

> *if*   X = the number of requests in a queue *and*
>   X is high
> *then*   RT > 3

> *if*   Y = NFS server *and*
>   X = number of requests in queue of Y *and*
>   X is high
> *then*   RT > 3

Interestingly, those rules are a generalization of the results of the ASM algorithms because Server 11 and five others are NFS servers.

Tilde was presented with the same table and induced the decision tree in Figure 5.26.

The output of Tilde is in agreement with the output of Progol. The joint parameters "X = NFS Server" and "queued(X)" have the greatest impact on RT > 3. Both Tracer and Server 11 are instances of NFS Server. Note that in the lower path where "queued(X) = low" for the class "NFS server," Tilde splits on "CPU_load(X)" for Server 11. One can interpret that to mean that high activity on Server 11 is the main influence on RT > 3.

Recall that from the ASM propositional experiments we concluded that memory problems or application overloading on Server 11 were the main influences on the SPT service. Here we see something similar. When

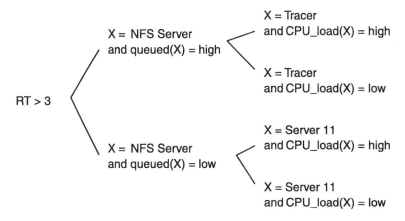

Figure 5.26 The decision tree produced by Tilde (SPT).

NFS activity on Server 11 is low, high CPU activity on Server 11 is the main bottleneck.

We can identify the situation with Server 11 as a swapping problem. The machine has low NFS activity but is swapping memory, causing high CPU activity. Again, the conclusion is that Server 11 needs more memory or that the number of applications on this server should be restricted.

Discussion: Data mining in enterprise management

The experiments in applying data mining methods to archived data are encouraging, and have produced results that are meaningful in a real-world example.

The task at KLM is a common one: to understand the causes that affect the behavior of SLA performance metrics. Clearly, if such causes can be discovered, we are halfway toward making performance better and thus more likely to meet SLA commitments.

A number of other tasks are amenable to data mining methods, each one deserving some old-fashioned brainstorming, discussion, and implementation.

▶ Identify the common characteristics of applications or BPs that appear to perform well, in contrast to those which perform poorly.

▶ Predict which applications or processes are likely to perform poorly in the future.

▶ Identify security breaches in network usage.

▶ Understand what networking services and applications are commonly used together.

▶ Reveal the difference between an application or process this month versus last month.

▶ Identify the causes of faults in historical data sets.

▶ Evaluate new network configurations.

The preliminary experiments and the KLM case study presented in this section suggest that the application of data mining techniques to the domain of enterprise management can offer insight into the dynamics of service performance.

Summary

This chapter examined eight problems in SLM. First, it looked at the problem of event correlation over the enterprise. We examined five approaches to the event correlation problem in the industry:

▶ Rule-based reasoning (Tivoli TME, BMC Patrol, Micromuse Net-Cool, Intellitactics Network Security Manager, and others);

▶ Model-based reasoning (Cabletron Spectrum);

▶ State transition graphs (Seagate NerveCenter);

▶ Codebooks (SMARTS InCharge);

▶ Case-based reasoning (Cabletron SpectroRx).

Those methods were examined with respect to several important dimensions: knowledge representation, knowledge acquisition, scalability, learning, and adaptation.

Second, in exploring the semantic disparity problem, we saw that a service is viewed differently among users, business overseers, and computer specialists. It is hard to reconcile those views. There are three approaches to the semantic disparity problem: the user-centric approach, the happy-medium approach, and the techno-centric approach. We argued that the user-centric approach should be favored and investigated to the fullest.

Third, we examined the component-to-service mapping problem, which manifests itself when we take the techno-centric approach to the semantic disparity problem. We described standard procedures for mapping values of component parameters to higher level values of service parameters. Next, we outlined an advanced method, fuzzy logic, that appears to be especially suited for translating numeric values into higher level common sense language.

The fourth, fifth, and sixth problems discussed were the agent selection problem, the integration problem, and the scaling problem. We group these problems together because they enter into the construction of a design model for an SLM solution. We argued that a design model is posterior to an analysis model. An SLM analysis model describes an SLM solution without regard to environmental constraints such as available monitoring agents, integration methods, or scalability. When we transform the analysis model into a design model, we face those problems head-on.

Seventh, we examined the representation problem. We distinguished between component-centric and process-centric representations of a service. The former is more common in the industry, while the latter is a challenging research problem.

Eight, we discussed the complexity problem. We argued that even though we do our best when constructing analysis, design, and testing models for services before we put them into production, there always seem to be unforeseen glitches. The complexity problem is that of identifying components in the enterprise that affect the performance of a service. Because a service can be affected by a subtle interplay among multiple components, this is a hard problem.

The case study at KLM Airlines addressed the complexity problem. The case study uses data mining methods to discover the influences on a service. We described three lab experiments that led up to the real-world experiment at KLM Airlines. Then we discussed the results of data mining methods applied to a large amount of enterprise performance data collected by monitoring agents at KLM.

Exercises and discussion questions

1. Section 5.1 discussed five approaches to the event correlation problem. Other approaches include neural networks, fuzzy logic, Petri nets, and genetic algorithms. Pick one approach and research it. Outline how the approach might solve the event correlation problem. How does it rate with respect to the considerations discussed in Section 5.1?

2. Another research question: What is a mobile agent? How does the mobile agent paradigm figure into event correlation? Into enterprise management in general?

3. Sections 5.2 and 5.3 described availability and response time as user-centric measures of a service. Consider an alternative measurement, "user happiness." Suppose users had a "user happiness button" on their screens and they are instructed to press the button whenever they are not happy with the service provided. What are the advantages and disadvantages of this approach?

4. Consider the discussion in Section 5.3 of a component-to-service mapping function for availability. At the end of the discussion, you were invited to think further about other possible problems with the function. Are there other problems? If so, how can their solutions be incorporated into the function? Is the function exactly what we want? Explain.

5. This chapter argued that the user-centric approach to service measurement should be investigated to the fullest before the happy-medium approach or the techno-centric approach is considered. Do you agree or disagree? Explain.

Further studies

Work on event correlation goes back to the early 1980s and today is stronger than ever. Listed here are several references to old and current work in event correlation. The component-to-service mapping problem is rather recent; most of the papers listed are from 1999. A good sampling of fuzzy logic papers is provided (a vast amount of other work in this area exists). Likewise, there is a good sampling of data mining work, although one will find lots more in the literature.

We wish to highlight a few references that were promised in the body of the chapter.

Russell and Norvig's textbook *Artificial Intelligence: A Modern Approach* is a good source for learning more about hard challenges in RBR research (and AI in general). The book also contains other topics that are relevant to this chapter. A very good paper on event correlation is Katker and Paterok's "Fault Isolation and Event Correlation for Integrated Fault Management." It contains a good (albeit short) comparison of most of the event correlation methods described in Section 5.1. One can learn about the coding algorithm in the codebook approach in the paper by Kliger et al., "A Coding Approach to Event Correlation," and the references cited therein.

Zadeh's paper, "Outline of a New Approach to the Analysis of Complex Systems and Decision Processes," is considered to be the seminal work on fuzzy logic. Lee's later paper, "Fuzzy Logic in Control Systems: Fuzzy Logic Controller" (Parts I and II), is an excellent discussion of standard fuzzy techniques.

The third lab experiment in the case study is described in the paper by Ibraheem et al., "Capturing a Qualitative Model of Network Performance and Predicting Behavior." The case study at KLM is discussed in the paper by Knobbe et al., "Experiments With Data Mining in Enterprise Management."

Select bibliography

Event correlation

Cronk, R., P. Callahan, and L. Berstein. "Rule-Based Expert Systems for Network Management and Operations: An Introduction." *IEEE Network Magazine*, Vol. 5, No. 4, Sept. 1988.

Frey, J., and L. Lewis. "Multi-Level Reasoning for Managing Distributed Enterprises and Their Networks." In *Integrated Network Management V.* London: Chapman and Hall, 1997.

Frontini, M., J. Griffin, and S. Towers. "A Knowledge-Based System for Fault Localization in Wide Area Networks." In I. Krishnan and W. Zimmer (eds), *Integrated Network Management II*. Amsterdam, North Holland: Elsevier Science Publishers, 1991.

Hasan, M., B. Sugla, and R. Viswanathan. "A Conceptual Framework for Network Management Event Correlation and Filtering Systems." In *Integrated Network Management VI*. New York: IEEE Press, 1999.

Houck, K., S. Calo, and A. Finkel. "Towards a Practical Alarm Correlation System." In *Integrated Network Management IV*. London: Chapman and Hall, 1995.

Jakobson, G., and M. Weissman. "Real-Time Telecommunication Management: Extending Event Correlation with Temporal Constraints." In *Integrated Network Management IV*. London: Chapman and Hall, 1995.

Katker, S., and M. Paterok. "Fault Isolation and Event Correlation for Integrated Fault Management." In *Integrated Network Management V*. London: Chapman and Hall, 1997.

Kliger, S., S. Yemini, Y. Yemini, D. Ohsie, and S. Stolfo. "A Coding Approach to Event Correlation." In *Integrated Network Management IV*. London: Chapman and Hall, 1995.

Lewis, L. "A Case-Based Reasoning Approach to the Management of Faults in Communications Networks." In *Proc. IEEE INFOCOM'93*, New York: IEEE Press, 1993.

Lewis, L. "A Case-Based Reasoning Approach to the Resolution of Faults in Communications Networks." In H.-G. Hegering and Y. Yemini (eds.), *Integrated Network Management*. Amsterdam, North Holland: Elsevier Science Publishers, 1993.

Lewis, L. *Managing Computer Networks: A Case-Based Reasoning Approach.* Norwood, MA: Artech House, 1995.

Liu, G., A. Mok, and E. Yang. "Composite Events for Network Event Correlation." In *Integrated Network Management VI*. New York: IEEE Press, 1999.

Mayer, A., S. Kliger, D. Ohsie, and S. Yemini. "Event Modeling with the MODEL Language." In *Integrated Network Management V.* London: Chapman and Hall, 1997.

Nygate, Y., "Event Correlation using Rule and Object Based Techniques." In *Integrated Network Management IV.* London: Chapman and Hall, 1995.

Russell, S., and P. Norvig. *Artificial Intelligence: A Modern Approach.* Englewood Cliffs, NJ: Prentice Hall, 1995.

Schroder, J., and W. Schodl. "A Modular Knowledge Base for Local Area Network Diagnosis." In I . Krishnan and W. Zimmer (eds), *Integrated Network Management II.* Amsterdam, North Holland: Elsevier Science Publishers, 1991.

Sutter, M., and P. Zeldin. "Designing Expert Systems for Real-Time Diagnosis of Self-Correcting Networks." *IEEE Network Magazine*, Vol. 5, No. 4, Sept. 1988

Vesonder, G., et al. "ACE: An Expert System for Telephone Cable Maintenance." *Proc. 8th International Joint Conf. on Artificial Intelligence*, Karlsruhe, Germany. Aug. 8–12, 1983

Williams, T., P. Orgren, and C. Smith. "Diagnosis of Multiple Faults in a Nationwide Communications Network. *Proc. 8th International Joint Conf. on Artificial Intelligence.* Karlsruhe, Germany. Aug. 8–12, 1983, pp. 179–181.

Wright, J., J. Zielinski, and E. Horton. "Expert Systems Development: The ACE System." In J. Liebowitz (ed), *Expert System Applications to Telecommunications.* New York: John Wiley and Sons, 1988.

Component-to-service mapping

Bhoj, P., S. Singhal, and S. Chutani. "SLA Management in Federated Environments." In M. Sloman and S. Mazumdar (eds), *Integrated Network Management VII.* New York: IEEE Publications, 1999.

Dreo Rodosek, G., and T. Kaiser. "Determining the Availability of Distributed Applications." In A. Lazar, R. Saracco, and R. Stodler (eds), *Integrated Network Management V.* London: Chapman and Hall, 1997.

Dreo Rodosek, G., T. Kaiser, and R. Rodosek. "A CSP Approach to IT Service Management (Poster Session)." In M. Sloman and S. Mazumdar (eds), *Integrated Network Management VII.* New York: IEEE Publications, 1999.

Frolund, S., M. Jain, and J. Pruyne. "SoLOMon: Monitoring End-User Service Levels." In M. Sloman and S. Mazumdar (eds), *Integrated Network Management VII.* New York: IEEE Publications, 1999.

Hegering, H.-G., S. Abeck, and R. Weis. "A Corporate Operation Framework for Network Service Management." *IEEE Communications Magazine,* Jan. 1996.

Hellerstein, J.-L., F. Zhang, and P. Shahabuddin. "An Approach to Predictive Detection for Service Management." In M. Sloman and S. Mazumdar (eds), *Integrated Network Management VII.* New York: IEEE Publications, 1999.

Kuepper, A., C. Popien, and B. Meyer. "Service Management Using Up-to-Date quality properties." *Proc. IFIP/IEEE International Conf. on Distributed Platforms: Client/Server and Beyond: DCE, CORBA, ODP and Advanced Distributed Applications.* London: Chapman and Hall, 1996.

Ramanathan, S., and C. Darst. "Measurement and Management of Internet Services Using HP Firehunter." In M. Sloman and S. Mazumdar (eds), *Integrated Network Management VII.* New York: IEEE Publications, 1999.

Fuzzy logic

Cox, E. "Fuzzy Fundamentals." *IEEE Spectrum,* Oct. 1992.

Lee, C. "Fuzzy Logic in Control Systems: Fuzzy Logic Controller" (Parts I and II). *IEEE Trans. on Systems, Man, and Cybernetics,* Vol. 20, No. 2, Mar./Apr. 1990.

Lewis, L. "A Fuzzy Logic Representation of Knowledge for Detecting/Correcting Network Performance Deficiencies." In I. Frisch, M. Malek, and S. Panwar (eds), *Network Management and Control,* Vol. 2. New York: Plenum Press, 1994.

Zadeh, L. "Outline of a New Approach to the Analysis of Complex Systems and Decision Processes." *IEEE Trans. on Systems, Man, and Cybernetics,* SMC-3, 1973.

Data mining

Adriaans, P.W., and R. Zantinge, *Data Mining.* Reading, MA: Addison-Wesley, 1996.

Fayyad, U.M., et al. (eds), *Advances in Knowledge Discovery and Data Mining,* Cambridge, MA: AAAI Press/MIT Press, 1996.

Ibraheem, S., M. Kokar, and L. Lewis. "Capturing a Qualitative Model of Network Performance and Predicting Behavior." *J. Network and System Management,* Vol. 6, No. 4, 1998.

Knobbe, A., D. van der Wallen, and L. Lewis. "Experiments With Data Mining in Enterprise Management." In *Integrated Network Management VI.* New York: IEEE Press, 1999.

Lavra, N., and S. Deroski. *Inductive Logic Programming, Techniques and Applications.* Hellis Horwood, 1994.

Quinlan, J. R. *C4.5: Programs for Machine Learning.* Morgan Kaufman, 1992.

de Raedt, L. (ed), *Advances in Inductive Logic Programming.* Amsterdam: IOS Press, 1996.

Venter, F. J. *The Use of Lattice-Based Knowledge Discovery to Determine Dependencies Between Network Management Parameters.* Master's thesis, University of Pretoria, South Africa, Oct. 1997.

Zantinge, R., and P. W. Adriaans, *Managing Client/Server Communications.* Reading, MA: Addison-Wesley, 1996.

In which we look at SLM in relation to electronic commerce over the Internet.

In this chapter:

▶ What is electronic commerce?

▶ Burdens on suppliers of electronic commerce

▶ SLM and electronic commerce

▶ Case study: Windward Consulting Group

SLM and electronic commerce

Previous chapters covered a lot of ground regarding SLM. We began with an intuitive understanding of SLM in Chapter 1, moved on to concepts and definitions in Chapter 2, SLM methodology in Chapter 3, SLM architecture in Chapter 4, and finally special topics and challenges in Chapter 5. Although we considered case studies of several vertical markets to illustrate the concepts in each chapter, the general discussion has been more of a broad horizontal nature.

At this point, it would be instructive to discuss detailed applications of our SLM guidelines to a variety of vertical markets and their special requirements, for example, health care, government systems, distance learning, telemedicine, banking, materials production, and so on. However, we will have to be content with a detailed discussion of just one market—electronic commerce.

We will discuss the requirements and the architecture of a system to manage distributed Web server farms that support electronic commerce for multiple industries. Clearly, SLM is crucial in that arena.

The electronic commerce market is interesting because its SLM dimension is doubly hard. The industries that employ suppliers of electronic commerce have to be assured that their customers can engage their Web sites without troubles or headaches. Otherwise, the industries run the risk of losing business (consequently, their suppliers also run the risk of losing business). Thus, the suppliers of electronic commerce have to answer to both industry and the industry's customer base.

The case study in this chapter is based on work performed by the Windward Consulting Group (United States). The author is grateful to Windward for sharing their expertise in the electronic commerce market. Windward interviewed and studied several electronic commerce providers from which they compiled a comprehensive set of requirements for end-to-end management of the operational infrastructure, including technical and business operations. From those requirements, we develop a scalable enterprise management architecture to support the expected growth of electronic commerce during the first quarter of the twenty-first century.

6.1   What is electronic commerce?

The author knows that the reader knows about electronic commerce (EC). But for the record, let us define it as follows:

> Electronic commerce is the process of buying and selling goods to consumers electronically and from company to company through computerized business transactions.

Consider EC from a historical perspective:

In 1872, Montgomery Ward & Co. launched the first mail-order catalog in the United States. The catalog offered home goods to thousands of people living in small towns and on farms. The idea flourished, and many other companies followed the trend. Throughout the first three quarters of the twentieth century, there were a number of thick catalogs around most people's houses (especially if one lived in the country). Some people probably have fond memories of the different purposes to which those catalogs were put.

In the mid-1990s, an entrepreneur created an electronic Catalog Site on the Internet to take advantage of global commerce and the information revolution. The Catalog Site is an electronic mall for companies selling their merchandise through catalogs.

The Catalog Site lists 200+ global companies that sell their wares through either on-line or traditional hard-copy catalogs. Visitors can scroll through the list of tenants or search for the goods they want by product type. When they click to select a company in which they are interested, they are taken to an electronic storefront offering telephone and fax numbers, store hours, types of payments accepted, and other essential information. They can link to the company's Web site to see its electronic display of goods, preview paper catalogs, find out what items are on sale, order gift certificates, and sign up for a biweekly e-mail newsletter.

In the same way that the hard-copy catalog medium for domestic trade became prominent as we entered the twentieth century, the electronic medium for international trade is becoming prominent as we enter the twenty-first century. Print shops flourished in the twentieth century to support the catalog medium of trade; now we see EC providers flourishing in the twenty-first century. We can imagine the burdens that printer shops had to deal with back then with their new opportunities. Similarly, we should be able to appreciate burdens that EC providers are dealing with now.

6.2 Burdens on suppliers of electronic commerce

EC is growing at a very fast rate. Industries increasingly depend on the Web as a way to advertise their goods. The current trend of industry is to employ EC providers to develop and maintain their Web sites. Good examples of electronic commerce providers are Intermedia Digex and GTE Internetworking in the United States.

For many EC suppliers, what were only recently a couple hundred Web-hosting systems in a central location have grown to several thousand systems partitioned into Web server farms. For example, an EC provider may have Web server farms located in many strategic locations around the globe. A crude analogy would be that the Web servers on the farms are multiplying like rabbits, and the rabbits are migrating to other breeding places around the globe. For that reason, it is clear that asset, change, and growth management are key concerns of EC suppliers.

The consumers of EC services are twofold. On the one hand, the consumers are the industries that employ EC firms to develop and maintain their Web sites. On the other, the consumers are the people who navigate to industry Web sites via TV and radio advertisements, word of mouth, or Web searches.

Industries are placing significant portions of their business revenues and customer relationships into the hands of their EC service providers as they push more products and services over the Internet. Consequently, these industries are demanding stringent performance and availability guarantees on Web-based transactions, with significant penalties for nonperformance.

EC suppliers work with industries to develop EC requirements, for which the EC supplier provides installation, operations, maintenance, and support services of various levels based on SLAs. That means establishing the capability to monitor SLA compliance in near real-time, to proactively avoid problems and to prioritize recovery efforts when malfunctions cause violations of SLAs.

The burden of managing SLAs has been transformed to a large extent on the management of the networks, systems, and applications within the EC enterprise on which the Web-hosting services depend. For example, at the network level, the GTE Internetworking Web-hosting services rest on integrated SONET OC-192 rings, Frame Relay, and an ATM-layered architecture. At the systems level, Web-hosting platforms include UNIX and NT servers. At the applications level, products include Microsoft IIS server, Cybercash payment system, iCat's storefront creation software, Web Advantage hosting services, and GTE's Commerce Reach system.

As a result of the complexity in EC, suppliers of EC have embarked on projects to develop an enterprise management system that can (1) manage the enterprise infrastructure of their EC business, (2) represent and monitor SLAs, and (3) scale with expected increasing demands of EC.

6.3 SLM and electronic commerce

The primary products sold and supported by EC providers are configuration, deployment, and management of Internet Web-hosting services for commercial industries. These services can span multiple servers and multiple Web-hosting farms.

Service offerings often are sold as bundles. For example, a bundle may include managed firewall services and dedicated network access. The bundling of services may require coordination across organizations and thus increases the complexity of SLA management.

All services are sold with varying SLAs, which quantify systems performance, service availability, backup completions and restore times, and problem resolution metrics. SLAs typically provide financial incentives for exceeding requirements and penalties for failing to meet performance objectives.

The primary service parameters that the industry and the public are interested in are:

▶ *Availability.* They want their Web sites to be available at all times.

▶ *Quick response time.* They do not want their customers to experience unbearable slowness when retrieving information or moving around screens.

▶ *Security.* They want to be assured that no intruders (e.g., competitors) can sabotage their Web sites, and they want to be assured of secure transactions with respect to personal information such as credit card numbers.

▶ *Integrity.* They want the words and the pictures on their screens to be clear and enticing, and they want the information to be accurate and up-to-date.

Performance metrics for SLAs typically are based on Web availability to the Internet and measurements of site access times. Availability usually is defined as the total minutes that the Web server is actually available to the public. Access time usually is measured on a regional basis using benchmarking methods.

With recent networking technology such as packet marking, differential services, and switched networks, EC providers are able to offer different levels of service in each of those categories, and customers can choose their preferences. If customers want 100% availability, optimal response time, and maximal security and integrity, then they pay more. Otherwise, they pay less.

Figure 6.1 shows a sample form for specifying an SLA. The form uses a calendar, and each day of the month is divided into four six-hour blocks. A customer marks the blocks with certain grades of availability (90–100%), certain grades of response time (2–5 sec), and certain grades of security (low, medium, or high). There is a default category at the bottom of the form that applies unless the calendar is marked otherwise.

The EC provider is able to set prices. For example, during the month of December, 100% availability costs X dollars, 99% costs Y dollars, and so on. During a major TV event, the provider may want to increase the price.

A customer manipulates the calendar with respect to various service grades to see what the cost will be. The total cost is updated as the customer marks the calendar. The customer can send the contract to the EC provider for approval or cancel out.

The monthly bill depends on the extent to which the service agreement is met or violated. For example, 100% availability is hard to achieve. If an agreement specifies 100% availability for an entire month and the provider demonstrates that the server has been available 100%, the supplier receives a bonus of X dollars in addition to the regular fee. If the agreement is not met, the provider is penalized. The provider publicizes such policies, and they are listed in the "Policies" section of the agreement.

6.4 Case study: Windward Consulting Group

This case study is based on work performed by the Windward Consulting Group (United States). Windward interviewed and studied several electronic commerce providers. They looked at the organizational structure

Service Agreement with XYZ Server Farm

Name
Address
Phone
Email

Policies

Availability   ___ (select 90 – 100 %) $___

Response Time   ___ (select 2 – 5 sec) $___

Security   ___ (select high-   med-low) $___

Integrity   ___ (select high-   med-low) $___

Total: $____

Go Back   *(Month)*   *Go Forward*

Default: Availability ____ Response time ____ Security ____ Integrity____

Send   Cancel

Figure 6.1   Sample form for an SLA.

of EC firms, their technical requirements, and their wish lists. From that they compiled a set of requirements for a scalable enterprise management architecture to support the expected growth of electronic commerce during the first quarter of the twenty-first century. We will examine these requirements and propose conceptual and physical architectures for an EC enterprise management system.

Organizational structure of EC businesses

As a result of the growth of EC and the technical operations involved, EC organizations and personnel likewise have increased significantly. What was once a small cohesive organization has transformed into a large enterprise. That growth has put extra stress on both business and technical operations within the organization. The management solutions and procedures that once worked fine are beginning to buckle under the stress.

The EC enterprise is fairly traditional in its organizational structure. Figure 6.2 shows a typical organizational structure.

Two of the departments, Technical Operations and Client Services, are especially interested in the prospect of a robust enterprise management system (EMS) because they are involved directly with the deployment and maintenance of Web-hosting services, problem management, and customer support. In addition, as we will see, other departments in the organization also benefit from the EMS.

Figure 6.2 Organizational structure of an EC enterprise.

A key ingredient driving the design and architecture of the EMS is the mapping of information flows across the organization. The EMS is expected to assist the organization by collecting and processing large amounts of systems and operational data, transforming them into information and knowledge, and presenting them to appropriate business departments within the organization.

For example, interviews conducted with EC organizations indicate that there needs to be much better sharing of data between the Technical Operations unit and other units in the business. The Sales, Marketing, and Strategic Planning units require access to historical data regarding system performance, possible service configurations, SLA compliance reports, and event exception reporting. Conversely, the Technical Operations unit needs to access data collected by the Sales and Marketing unit to plan for expansions to the technical infrastructure. If that could be achieved, the Technical Operations unit would be better aligned with the business units and vice versa.

As another example of information flow, the Client Services unit, particularly the people who run the help desk and oversee problem management, need to be cognizant of operations information collected by the EMS. If that could be achieved, trouble tickets and complaints logged by consumers could be correlated with operational data such as outages and traffic bottlenecks. Thus, there is a requirement to create an interface between the help desk and the EMS to allow information to be shared between the Technical Operations unit and the Client Services unit. Such sharing of data would go a long way toward efficient problem management.

EMS requirements for the EC business

This section is a bit detailed, but it is good to have at least one detailed discussion of the EMS requirements of some vertical market. It is likely that some of the discussion in this section will carry over to other domains.

Windward identifies the following crucial features for an integrated EMS that can manage and scale for EC:

▶ Monitoring;

▶ Event management;

- Reporting;

- Configuration, asset, and change management;

- Software distribution;

- Problem management and automated fault management;

- Trend and performance analysis;

- Security management;

- SLA management;

- EMS console.

Monitoring

The requirement in this category focuses on the ability of the EMS to monitor the various systems, network devices, and software applications for real-time display and historical reporting. The EMS should provide visibility into operational parameters that provide operations staff with meaningful information for maintaining network and systems availability and performance.

The key monitoring requirements for computer systems are these:

- Provide cross-platform support for NT, Solaris, SunOS;

- Monitor key performance indicators (CPU, memory, disk, I/O, NIC cards);

- Monitor key components (power supply, unit temperature);

- Monitor systems and error logs for specific data and events;

- Monitor selected processes for state (up/down);

- Monitor resource utilization of selected processes (CPU/memory/I/O);

- Monitor security events.

The key monitoring requirements for networks are these:

- Monitor devices from major network equipment vendors;

- Provide access to MIB II variables supported by the network hardware vendor;

- Provide access to vendor management extensions built into the hardware;

- Receive and process all network device traps;

- Detect potential security breaches and denial of service attacks and generate appropriate events.

The key monitoring requirements for applications are these:

- Monitor industry-standard databases (Oracle, Sybase, MS SQL Server);

- Monitor key database performance items (table size and growth rate, redo log size and growth rate, resource utilization, and connectivity);

- Monitor Web server engines for hit rates and error conditions;

- Monitor system backups for proper execution and completion.

The key monitoring requirements for end-to-end functionality and connectivity are these:

- Monitor TCP ports from a remote server and report on functional status (e.g., HTTP, Sendmail, DNS, FTP);

- Execute a Web page transaction from a remote client and monitor performance and content integrity.

The key monitoring requirements for facilities are these:

- Monitor computer room environmental (temperature, humidity, power, water);

- Monitor uninterrupted power source (UPS), generators, and air handlers for status and alarms.

Event management

It is the job of event management to take information from the monitoring agents, log it, filter it, correlate it, and determine what actions and notifications, if any, need to take place. Technical Operations needs visibility into the events of the thousands of managed items in the EC enterprise.

The EMS should allow the Technical Operations staff to become proactive in preventing service interruptions by identifying and responding to low-impact events that may be precursors to a more serious event that would cause a service outage.

The key event management requirements are as follows:

- Correlate multiple events to assist in root cause determination;
- Send notification via multiple methods (e.g., e-mail, pager, operations console);
- Identify events by client by correlating the originating device to client;
- Provide automated responses to events;
- Maintain a unified repository of historical events for analysis;
- Set customized thresholds based on user;
- Automatically open a trouble ticket;
- Escalate alarms based on time and persistence;
- Accept and process non-SNMP events;
- Direct specific events to specific consoles;
- Provide fault tolerance so events are not lost due to EMS system failure;
- Suppress nuisance events from the event stream.

Reporting

Operational data obtained by the monitoring agents must be transformed into management information. The EMS has to make the appropriate data available as specialized reports to support the needs of both the business and technical operations within the organization.

The classes of reports are these:

- SLA compliance/exception reports;
- Daily event summary reports;

- Asset and configuration reports;
- Performance exception reports;
- Security event reports;
- Performance trending reports;
- Alarm and event trending reports;
- Outage reports;
- Utilization and capacity reports;
- Various reports on trouble-ticket status and closure rates;
- Backup operations and tape management reports;
- Billing reports;
- Specialized event-driven reports based on an event in the enterprise, for example, some subset of the other reports listed.

Configuration, asset, and change management

Configuration management is the discipline by which changes to a known environment are performed according to prescribed processes. The processes require testing and documentation prior to implementation. The documentation requirements include configuration records, asset records, date and time of changes, appropriate signoffs, updates to policies and procedures, and so on.

The EMS should be used to detect and report unexpected changes, as well as to collect comprehensive information about the state of current configurations. Additionally, the EMS is used to provide the means to roll back to previous configurations.

Configuration and asset management procedures apply across the organization. Unplanned or uncoordinated changes made in one area of the environment might have unwanted results in another. Additionally, configuration and topology changes in the environment can change the context and meaning of information used by event management, thereby undermining the ability of the event management system to provide an accurate view of the condition of the enterprise.

Change management is probably the most essential process in the enterprise. Many people argue that if a company cannot manage both planned and unexpected changes in enterprise structure, then it cannot manage anything else. For example, suppose a service depends on a database server, where management data about the server and other components are mapped into the overall health of the service. Now imagine that a new server is installed and the database on the old server is transferred to the new server. The old server will be used for something else. An administrator must remember to redefine the service to make sure it includes the new server. Otherwise, the definition of the service is at odds with the change made to the enterprise structure. That simple fault could propagate to other faults in the overall management system.

To make matters worse, consider that some applications dynamically allocate their own resources. Thus, hard-coded service definitions that depend on those resources easily can become unsynchronized. Clearly, change management is a nontrivial task for an EMS.

The key configuration, asset, and change management requirements are these:

▶ Know the inventory of hardware and software configurations of each managed system in the enterprise;

▶ Support multiple platforms and operating systems (NT, Solaris, Sun OS, and other major UNIX flavors);

▶ Retain versioned copies of key configuration files for each managed object;

▶ Detect changes in the baseline configuration and immediately notify operations of the detected change;

▶ Create profiles for noncommercial software to allow for appropriate identification;

▶ Identify and track software to the version level;

▶ Discover new hardware that is introduced into the environment;

▶ Update inventory files and service definitions as changes occur in the enterprise structure.

Software distribution

The large number of systems that reside in server farms requires a robust method of managing the distribution of software when patches, upgrades, and new capabilities need to be installed. When combined with the configuration and asset management components of the EMS, automated software reduces the time and the effort to install software over a large system install base.

Software distribution installs and tracks software installations from a single console. Software can be targeted for distribution based on any combination of attributes tracked in the configuration management database. Additionally, the software distribution system provides alerts and status information on the success or failure of the installation process on each target machine.

The specific requirements for software distribution are these:

▶ Provide single-interface support for multiple platforms and operating systems (NT, Solaris, Sun OS, and other UNIX flavors);

▶ Send status and failure alerts to a console;

▶ Maintain configuration management for software delivery packages;

▶ Select target machines based on asset attributes maintained by asset management;

▶ Send an uninstall action to uninstall software.

Problem management and automated fault management

Problem management is a combined function of the people who operate the help desk and the people who maintain the enterprise. Trouble tickets may be generated manually by help desk operators or network administrators. In addition, trouble tickets may be generated automatically by the EMS itself. Hard alarms on the enterprise, such as outages and bottlenecks, should result in a high-severity trouble ticket. In that way, problems often can be resolved and managed before customers begin to experience them.

Automated fault management works in conjunction with the monitoring function and the event management function. It is an advanced capability that uses techniques in AI to recognize problems and take action with minimal human interaction. Thus, it reduces the need for operations staff to respond to routine events and allows them to focus on more complex issues.

Automated fault management requires a well-monitored environment in which the inventory of enterprise components and structure is kept up to date. As such, its success depends in large part on good configuration, asset, and change management practices.

The key requirements for problem management and automated fault management are these:

▶ Manual trouble-ticket generation;

▶ Automatic ticket generation;

▶ Problem tracking and escalation;

▶ Correlate events based on some identified AI paradigm;

▶ Transform events into alarms (both interdomain and intradomain alarms);

▶ Incorporate new correlation knowledge into the AI system;

▶ Offer solutions to identified problems;

▶ Provide explanations;

▶ Execute solutions under user supervision.

Trend and performance analysis

Trend and performance analysis improves the organization's ability to anticipate future needs and appropriately plan for them. Just as monitoring and event management provide the ability to respond quickly to real-time events and issues, trend and performance analysis provides the ability to proactively identify an area's operational and business risk. By identifying those risks early, they can be managed before they compromise the integrity of the operation or result in financial loss. Additionally, trend and performance analysis can reveal otherwise unseen opportunities to sell additional capabilities and services to existing clients.

The specific requirements for trend and performance analysis are these:

▶ Collect and trend key performance indicators for systems performance for each managed system (e.g., CPU, Memory, I/O, Swap, Disk, NIC);

- Collect and trend key network performance indicators for network devices (e.g., packet rate, packet loss, CPU memory, interface flapping);

- Collect and trend system availability and stability by customer and configuration;

- Trend SLA compliance levels and projected financial impact;

- Trend backup performance and data growth to ensure that backups and restores can be completed within allotted time windows;

- Trend problem resolution closure rates and times to close.

Security management

Effective and demonstrable security safeguards are strong product discriminators in the EC market. It is important that appropriate security measures are in place and that the integrity of security measures is well maintained to safeguard customer assets and operational integrity. The EMS ensures that common security threats are identified and may contain mechanisms to thwart threats when they occur.

The specific security requirements are these:

- Manage systems authentication from a single security database;

- Identify system files and user actions as security sensitive and receive alerts when those files are touched or intrusive actions are taken;

- Report alerts when security barriers have been challenged;

- Identify common network and systems security threats (e.g., password cracking, denial of service, port scanning), take action to thwart the attack, and send an alarm to the operations staff;

- Track customer administrative activities on their platform to ensure that operational policies have not been violated and platform integrity has not been compromised.

SLA management

SLA management is crucial for EC. Industries have to be convinced that their customers are not having problems accessing and using their Web sites. Further, decisions regarding operational activities, expenditures, and capital investment are measured against the existing and anticipated SLA compliance reports.

The specific requirements for SLA management are these:

▶ Report on service availability as determined by polling the service port (e.g., HTTP, FTP, SMTP POP3, SSL) at regular intervals to determine total time in minutes that service is not available during a given period of time;

▶ Capture and report file backup and restoration activities and status per machine for some given period;

▶ Calculate average data rate, in megabytes per hour, that files were restored from backup, where the start time is the time of the initial request and the stop time is the time that file restore was completed;

▶ Measure and report response time and problem fix time for each incident by the customer and determine if the SLA requirement was met based on the customer SLA;

▶ Capture and report, at defined SLA intervals, key systems perform- ance data (CPU, memory, disk space, and others as required) and present the maximum, minimum, and average utilization for each measure for a given period of time;

▶ Create consolidated SLA reports that encompass all elements of a customer's agreement;

▶ Capture and report network bandwidth utilization and other net- work and systems utilization data required for billing purposes;

▶ Monitor real-time events, make real-time SLA compliance risk assessments, and provide operations with a warning when an SLA metric is at risk of being violated.

EMS console

The EMS console supports all other functions. It provides the means to display various categories of information in support of each unit in the organization. Further, it provides the means to launch the tools required to manage specific parts of the enterprise. As such, the EMS console must be highly configurable and provide means to set permissions for different classes of users.

The specific requirements for the EMS console are these:

- Support alarm filtering;
- Provide both traditional GUI interfaces and Web interfaces;
- Object-oriented GUI (i.e., elements in the GUI are manipulated in the same manner regardless of type);
- Support for hierarchical topology maps;
- Provide GUI context information that can be passed on a command line to launch other applications;
- Programmable command execution buttons;
- Support multiple profiles and configurations by user logon;
- Provide logon security for controlling and limiting scope of activity for each operator;
- Provide appropriate security controls to allow client access to view their own systems.

Finally, given the large scale and fast growth of the EC server farms and the dynamic nature of the EC market, other general EMS requirements must be taken into consideration:

- It must be scalable to support multiple, geographically disperse operations centers.
- It must be capable of incrementally scaling to support an increasing number of managed systems and network devices while preserving performance and architectural integrity.
- It must provide the ability to view and monitor any managed object from any operations console.
- It must provide clean points of interface between functional systems and subsystems.

◗ It must be based on an object abstraction model.

◗ It must provide fault tolerance and redundancy.

A first-order analysis of the requirements

The core of the EMS is the event management feature (Figure 6.3). All submanagement systems (e.g., systems and network management, configuration management, software distribution, security) monitor their respective domains, perform their specialized work, and pass events of interest to the event management system for enterprisewide monitoring and control.

Figure 6.4 shows a conceptual architecture of an integrated EMS that takes us a step closer toward the system's structure.

Much of the discussion in previous chapters contributes to the architecture in Figure 6.4. To see that, thumb back through a series of figures in prior chapters:

◗ Figure 4.1 shows the basic EMS architecture.

◗ Figures 4.7, 4.8, and 4.9 show alternative schemes for data warehousing.

◗ Figure 4.10 shows an enhanced EMS architecture.

◗ Figure 4.12 shows an architecture for distributed domains (e.g., multiple Web server farms).

◗ Figure 5.5 shows distributed event correlation over multiple monitoring agents.

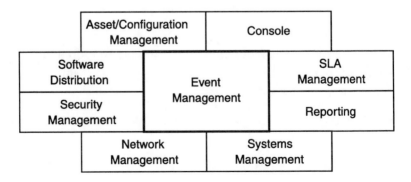

Figure 6.3   The prominence of event management.

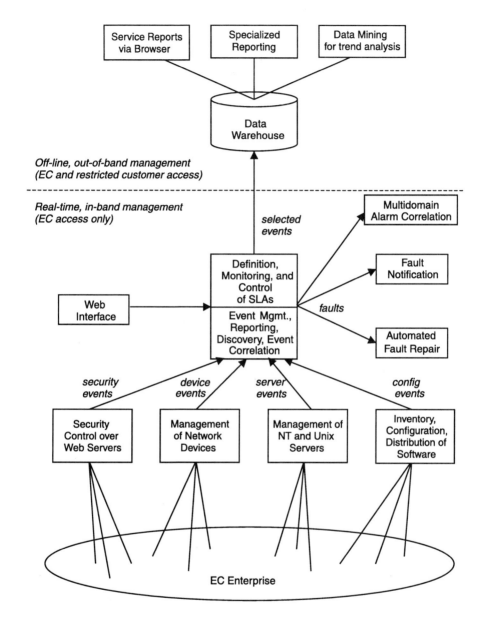

Figure 6.4 A conceptual EMS architecture for the EC business.

If we are satisfied that the conceptual architecture accommodates the requirements listed in the "EMS requirements for the EC business" section 6.4, we can take another step toward its realization. Figure 6.5 is an exact reflection of Figure 6.4, with the generic words in the boxes being replaced by candidate commercial products.

To the author's knowledge, a complete integration of the management tools in Figure 6.5 has not been implemented before, although one can find many integrations of subsets of the tools in the industry. We discussed some of the integrations in previous chapters. Here we won't explain just how the systems shown in Figure 6.5 actually work together. The reader is referred to a Spectrum integration manual and other literature in the Further Studies section for detail.

Several features of the integrated EMS architecture are worthy of note. First, the architecture reflects a network-centric management style. Compare Figure 4.2 and surrounding discussion. It is not hard to revise the architecture for system-centric and application-centric management styles.

Second, the architecture is plug-and-play. For example, GTE Internetworking employs Tivoli TME for systems monitoring, whereas other businesses might employ BMC Patrol, Metrix WinWatch, Platinum ServerVision, NetIQ's AppManager, or other management systems. Also, we have included the Clarify help desk as the means of fault notification, to illustrate the information flow between Technical Operations and Client Services in the EC organization. Of course there are other complementary examples of fault notification, including e-mail, paging, and traditional console alerting. In that regard, refer to Figure 3.5.

The consolidated enterprise console toward the center of Figure 6.5 provides a high-level view of the enterprise from a single console. For example, Figure 6.6 shows a simple Spectrum/ICS screen shot of a service decomposed into supporting network devices, computer systems, and applications. The three icons at the top of hierarchy (those that look like PacMan) represent services. We see that the IcsWebSite service is decomposed into two subservices (Internet Access and the Backbone), an HTTP Daemon, and a Web server. The light-colored icons represent low-level enterprise elements.

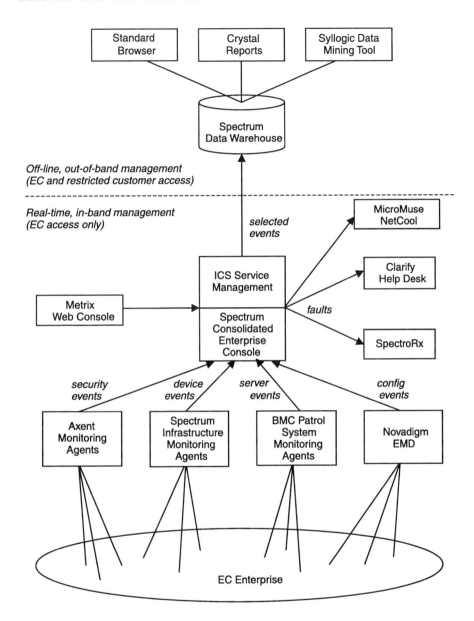

Figure 6.5 The physical architecture derived from the conceptual architecture.

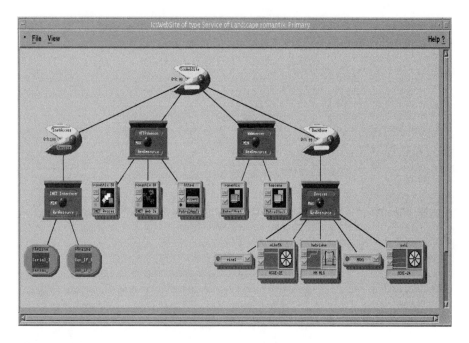

Figure 6.6   A Spectrum/ICS screen shot.

The pull-down View menu at the top of Figure 6.6 contains a list of possible views and actions (not shown) that can be executed from this console. In addition, the user can click on a particular component and see a list of actions specific to the component. For example, suppose a BMC patrol agent detects a fault in a server, which in turn affects the service. In this case, both icons might turn red, indicating an alarm. On the basis of the alarm, one can pick an action in the View menu that will generate a corresponding trouble ticket in the Clarify help desk, or it may pass surrounding information to SpectroRx in hopes of finding an explanation and repair procedure, or it may navigate to a detailed BMC view of the culprit server. The user can click on a service icon to view or modify the SLA for the service. Figure 6.7 shows the invocation of an SLA. Figures 6.8 and 6.9 show the invocation of various device views and the status of their ports. Figure 6.10 shows an event log for a particular device, and Figure 6.11 shows an alarm log for the entire

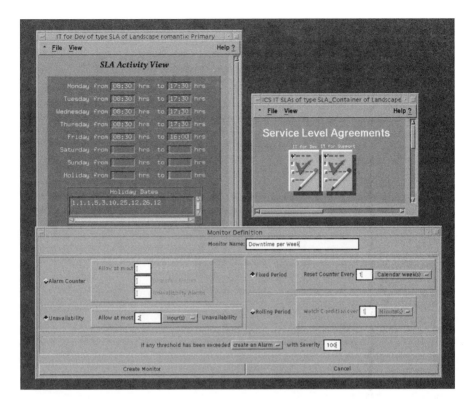

Figure 6.7 Invocation of a service level agreement.

enterprise. These views represent some of the screen shots available in the Spectrum EMS.

Clearly, there are other issues to consider as we begin to deploy the integrated EMS. For example, a strategic architecture and plan should be devised first. It would not be advisable to deploy the whole system all at once. Further, logistics, performance, and usability issues have to be considered. A good rule of thumb is to collocate selected subsets of client GUIs of each management application on various platforms within the EC organization based on information requirements. Systems analysts and architects are quite familiar with those concerns, so we will not go into them here.

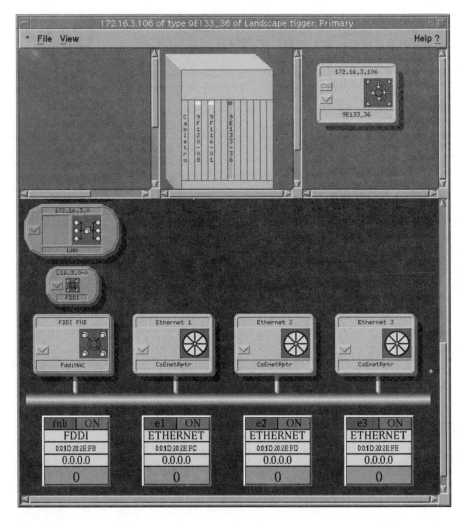

Figure 6.8   A component view showing port status.

Summary

This chapter looked at SLM in relation to electronic commerce. We described the burdens on EC providers in light of the growing demands for Web-hosting services. We looked at a detailed set of requirements based on studies by the Windward Consulting Group. From those requirements, we proposed a conceptual architecture, and from the conceptual architecture we derived a physical architecture.

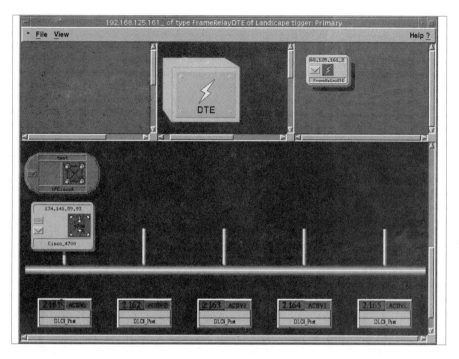

Figure 6.9 A component view showing port status.

Exercises and discussion questions

1. The conceptual architecture in Figure 6.4 suggests a unidirectional communication between the event management system and the submanagement systems. Security, device, server, and config events are passed to the central management system. Discuss scenarios in which a bidirectional communication is preferred. What classes of data would be involved in the reverse direction?

2. Explain the difference between in-band management and out-of-band management.

3. The Web site www.ics.de allows you to view live management of the Cabletron European Network. The integrated suite of management tools includes Spectrum, ICS Service Management, BMC Patrol, and the Metrix Web Console. Experiment with the system

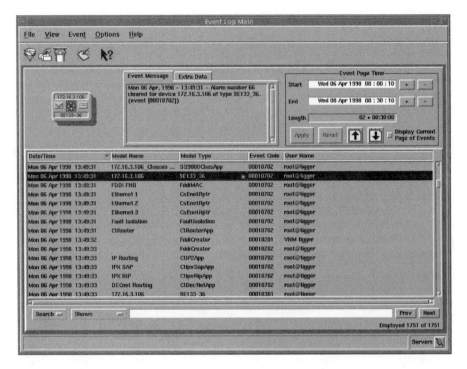

Figure 6.10   An event log for a particular device.

and provide two kinds of evaluation: (1) an evaluation from a consumer's point of view with respect to usability, content, and responsiveness and (2) an evaluation of the integrated management system from a technical point of view in light of the concepts in this book.

Further studies

For good examples of electronic commerce services and offerings, the reader can visit www.gte.com and www.digex.net. For further information on the work of the Windward Consulting group, visit www.windward.com.

Of course, industry Web sites are plentiful. Examples are www.barnesandnoble.com (for books), www.eBay.com (for auctions), and www.travelocity.com (for air fares).

Figure 6.11 An alarm log for the entire enterprise.

Regarding integration architectures and methods, the reader can visit the Web sites of the companies referenced in this chapter. Many vendors have their product manuals on the Web. For example, you can visit www.cabletron.com to get a copy of the *Spectrum Guide to Integrated Applications*. The guide discusses several generic classes of integrations, case studies, and samples of integration code that were used to construct the integrated architecture in Figure 6.5. To see methods for integrating EMSs and trouble tickets systems, see Lewis's book *Managing Computer Networks: A Case-Based Reasoning Approach* and Lewis and Dreo's paper "Extending Trouble Ticket Systems to Fault Diagnosis." Also see Dreo and Valta's paper "Using Master Tickets as a Storage for Problem Solving Expertise" and Penido et al.'s paper "An Automatic Fault Diagnosis and Correction System for Telecommunications Management."

The Web sites of vendors referenced in this chapter are as follows:
www.ics.de
www.micromuse.com

www.novadigm.com

www.bmc.com

www.axent.com

www.metrix.lu

www.seagatesoftware.com

www.syllogic.com

www.clarify.com

www.tivoli.com

www.platinum.com

www.netiq.com

Select bibliography

Communications of the ACM. Special Issue on "Multiagent Systems on the Net and Agents in E-Commerce." Vol. 42, No. 3, March 1999.

Dreo, G., and R. Valta. "Using Master Tickets as a Storage for Problem Solving Expertise" in *Integrated Network Management IV*. W. Zimmer and D. Zuckerman (eds), Amsterdam, North Holland: Elsevier Science Publishers, 1995.

Lewis, L. *Managing Computer Networks: A Case-Based Reasoning Approach.* Norwood, MA: Artech House, 1995.

Lewis, L., and G. Dreo. "Extending Trouble Ticket Systems to Fault Diagnosis," IEEE Network. Vol. 7, No. 6, 1993.

Penido, G., J. M. Nogueria, and C. Machado. "An Automatic Fault Diagnosis and Correction System for Telecommunications Management." In *Integrated Management VI.* M. Sloman, S. Mazumdar, and E. Lupu (eds). New York: IEEE Publishing, 1999.

In which we wrap up our book by looking at SLM in relation to things that are dear to most workers in modern business: working at home, quality of life, and productivity—all at the same time.

In this chapter:

▶ Information systems and modern business

▶ The SLM connection

▶ Business trends and challenges

▶ Why be interested in SLM?

SLM, modern business, and quality of life

It is well known that the success of a business depends on its efficient use of information, its organizational structure, its management procedures, its degree of internal communication, and its ability to predict and adapt to environmental changes in the marketplace.

Although this book is not about business science per se, the final chapter examines the connection between SLM and modern business. It is a good topic with which to round out the book.

First we compare the 1970s view of modern business with the 2000+ view. We see the roles that information systems (ISs) play in each view. Second, we look at the connection between SLM and ISs in general. The key is to understand that the notion of a "service" is synonymous with the "purpose" of an IS. Finally, we look at business trends and revisit the discussion "Why be interested in SLM?" from Chapter 1.

7.1 Information systems and modern business

The textbook definition of an IS is this:

> An information system is a set of interrelated components that collect, process, store, and disseminate information for purposes of decision making, coordination, control, analysis, and production of commodities in an organization.

The case studies in previous chapters illustrated specific examples of ISs, for example, electronic commerce in Chapter 6, tracking and tracing aircraft parts in Chapter 5, and intercontinental communication back in Chapter 1.

The goal of the case studies was to illustrate the particular SLM concepts discussed in each chapter. Here, however, we want to think about SLM as it relates to modern business.

The 1970s view of modern business

Discussions of modern business in 1970s literature centered on the topics of (1) the emergence of the global economy, (2) the information revolution, (3) horizontal (as opposed to vertical) management and organizational structures, and (4) the roles of ISs in all parts of the business.

ISs were seen to be pivotal instruments in business dynamics and business success. In the 1970s it was predicted that ISs would be used for the following tasks:

- Collection of data about the global marketplace;

- Analysis of data to infer trends and new opportunities and thereby establish new business goals;

- Internal communication with respect to organizational changes and assignment of duties;

- Support for the production of materials;

- External communication with consumers.

The predictions in the 1970s certainly turned out to be true. For example, modern industries that produce software depend on ISs so that multiple software engineers can write code for different pieces of the software at the same time. Quality assurance (QA) engineers use ISs to communicate with software engineers regarding bugs in the software and to alert managers if the bugs are not fixed in a timely fashion. Managers use ISs to remind software engineers of their duties, and so forth.

Other examples of uses of ISs in modern business are plentiful. These days, however, people are beginning to see other uses of ISs.

The 2000+ view of modern business

Let us continue the example from the previous section. Suppose a software engineer writes code in building A and deposits the code in a secure server in building B. A QA engineer in building C downloads the code from the server in B and begins testing it.

It is not hard to generalize that scenario: A software engineer writes code in country A and deposits the code in a secure server in country B. A QA engineer in country C downloads the code from the server in B and begins testing it.

And once more: A software engineer writes code at home and deposits the code in a secure server in building B. A QA engineer at home downloads the code from the server B and begins testing it.

One gets the idea. If a person is responsible for taking care of children or elderly relatives or has other personal obligations, the prospect of integrating those responsibilities with professional duties from the home is quite attractive. ISs are beginning to render that kind of life possible.

The example of the software industry is somewhat seductive, but we all realize that other things have to be taken into consideration for other kinds of industries. For example, the instinctive element of *human sympathy* (also known as the human touch) is often required to promote human collaboration toward a common goal, and the only way of experiencing human sympathy is to experience the real features and mannerisms of other humans in personal face-to-face meetings.

Modern-day ISs are beginning to support the element of human sympathy in electronic collaborations. A useful classification of the state of the science regarding electronic human sympathy, in increasing degree of sympathy, is this:

- *Teleconference*: Group conference via telephone or e-mail software;
- *Dataconference*: Teleconference in which participants also are able to edit and modify shared data files simultaneously;
- *Videoconference (VC)*: Dataconference in which participants also are able to see and hear each other over video screens.

A VC is about as close as we can come to real human sympathy. However, current research is in the middle of trying to realize it. Part of the problem is the development of VC technology to support it, and another part of the problem is understanding the purposes we want to achieve with VC technology.

Imagine yourself as a student in a classroom. An important part of the dynamics of the classroom experience is reading the facial expressions and mannerisms of the instructor and the students as the class unfolds. Sometimes the class is monopolized by a discussion between the instructor and some individual student, other times the class is monopolized by discussions among the students themselves. All this is part of the learning experience.

Now consider the remote classroom (also known as distance learning). Physically, you are alone at home, a hundred miles away from a major university, but logged into the class of a prominent professor. Other students also are alone in their homes. You and they have computer screens in front of you, through which you participate in the classroom experience.

The minimal experience of the virtual classroom is that you see and hear only the instructor. No questions can be asked, and after the lecture is over you log out.

A better experience is that you can hear the questions of other students. You also can ask questions. However, you cannot see your peers and they cannot see you.

The ultimate experience is the ability to switch at will from a view of the instructor to a view of any peer student or to a global view of all participants in the class.

From a technology perspective, the minimal experience requires one video stream from the university entering your home. The better experience requires additional synchronized teleconference streams from the students' homes into your home. And the ultimate experience requires yet additional synchronized video streams from the students' homes into your home.

The additional bandwidth and processing requirements increase significantly with respect to each kind of experience. The ultimate experience is not quite possible today, although it is the goal of distance learning and other types of services. This is a topic of current research, and the results will carry over to other industries. The Next Generation Internet (Internet2) shows much promise for making that happen.

7.2   The SLM connection

Consider again the classic definition of an IS, this time focusing on its general purposes:

> An information system is a set of interrelated components that collect, process, store, and disseminate information *for purposes of decision making, coordination, control, analysis, and production of commodities in an organization.*

It is not hard to see the connection between ISs and SLM. For example, compare the beginnings of our SLM conceptual framework in Chapter 2:

▶ A *business process* (BP) refers to some way in which a company coordinates and organizes work activities and information to produce a valuable commodity. A typical BP includes several general services in the process, and some of those services may depend on the business's enterprise network.

▶ An *enterprise network* consists of the following general categories of components: transmission devices, transmission lines among the devices, computer systems, and applications running on the computer systems.

▶ A *service* is a function that the enterprise network provides for the business. We can think of a service as an abstraction over and above the enterprise network. Alternatively, we can think of a service as an epiphenomenon, a phenomenon that arises in virtue of the structure and operation of the network.

If we equate "service" with "the purpose of an IS," we have a clear connection between SLM and business in general, and we see that business science encompasses SLM as a subdiscipline.

From a business perspective, the additional features that business science brings to our SLM discussions are methods for:

▶ Inferring a business process or business goal in reaction to the marketplace;

▶ Conceiving, designing, and implementing ISs to support new business opportunities and goals.

Those topics are clearly important, although they are outside our scope here. This book assumes that business goals and processes are understood, and that ISs have been identified that support the goals of the business. The primary subject of this book is about methods for ensuring that the ISs do in fact support the business.

When the business goals and processes have been identified, and when their supportive ISs are in the midst of design and development, then we are in the SLM space. The guidelines of SLM in this book come into play.

Unfortunately, it is not uncommon in industry that an IS is designed and put into production only to find that service levels are inadequate. Typically, problems occur with respect to usability, reliability, and responsiveness, and it takes a fair amount of time and frustration to work out the bugs. Considering SLM principles early on will help alleviate those problems.

Another problem with respect to the introduction of new ISs is the stress placed on organizational and management changes in the firm. Even if IS design and development teams pay heed to SLM principles, there still remains the problem of the organization and management having to adapt to them.

From a business perspective, that is an equally important challenge, although it is outside the scope of this book. Nonetheless, it is clear that adhering to SLM methods with respect to usability, reliability, and responsiveness will go a considerable way toward alleviating the problem.

In sum, we see that an IS is more than the technologies that support it, as shown in Figure 7.1. In general, we can say that an IS is a combined organizational-management-technological solution to a challenge posed by the vicissitudes in the marketplace.

7.3 Business trends and challenges

A trend in modern businesses is to organize the core business into various units, one of which usually is designated the IT unit. Figure 7.2 shows such an organization.

The IT unit is responsible for the development and maintenance of ISs across other functional units in the business. The IT unit may be owned by the core company, or it may be outsourced to a third-party contractor.

A second trend in modern business is to look at ISs and the flow of information across the organization in terms of the following business applications:

▶ Operation-level systems that support operational managers by keeping track of the elementary activities and transactions of the

Figure 7.1 Multiple dimensions of information systems.

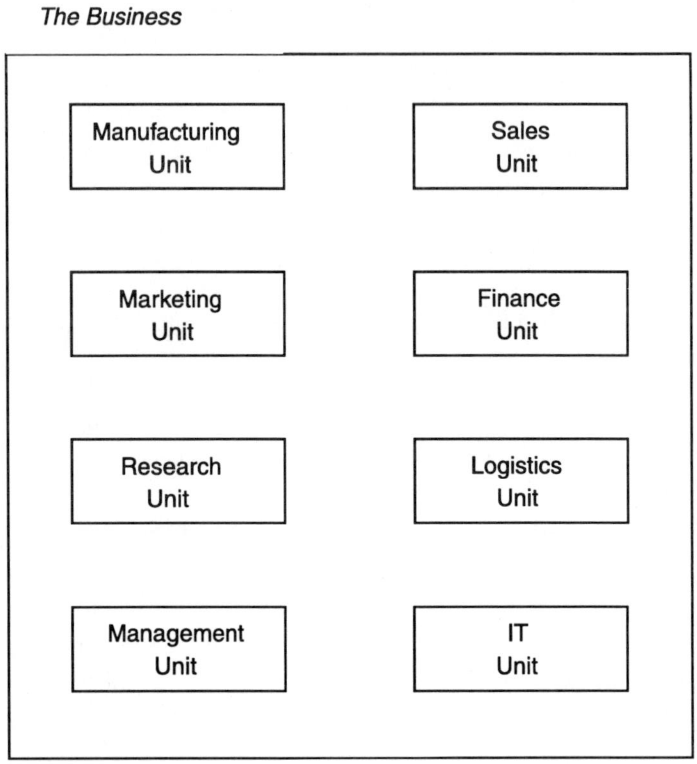

The Business

Figure 7.2 Example of a distributed business.

organization, such as sales, receipts, cash deposits, payroll, credit
decisions, and the flow of materials in a factory;

▶ Knowledge-level systems that support knowledge and data workers
in the organization by helping the firm discover, organize, and
integrate new knowledge into the business and to help the organization control the flow of paperwork;

▶ Management-level systems that support monitoring, control, decision making, and administrative activities of middle managers;

▶ Strategic-level systems that help senior management tackle and
address strategic issues and long-term trends, both in the firm and
in the external environment.

A third trend is to develop an information architecture for the organization. Figure 7.3 shows the information architecture of modern firms. It is the responsibility of the chief information officer and staff to understand how to arrange and coordinate the various computer technologies and business applications to meet the information needs at each level of their organization, as well as the needs of the organization as a whole.

Clearly, the construction of an information architecture for a firm is not a trivial task. Besides understanding IS technology and the SLM dimension of ISs, we must take into account the global marketplace and the nature of the firm's organization. Those challenges can be summed up as follows:

▶ *The strategic business challenge.* How can businesses use IT to design organizations that are competitive and effective?

▶ *The globalization challenge.* How can firms understand the business requirements and system requirements of global commerce?

▶ *The information architecture challenge.* How can organizations develop an information architecture that supports their business goals?

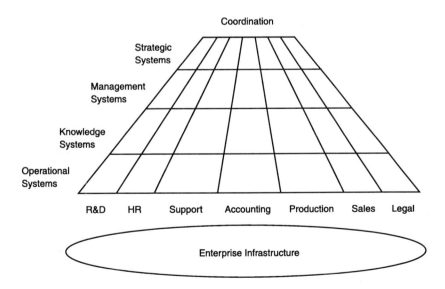

Figure 7.3 Information architecture of the organization.

▶ *The information systems investment challenge.* How can organizations determine the business value of ISs?

▶ *The usability challenge.* How can organizations design ISs that people can control and understand?

▶ *The SLM challenge.* How can an organization ensure that ISs continuously satisfy the needs for which they were designed?

We have not tried to answer the first five questions in this book. Our book has focused squarely on the last question, although it is not hard to see the effects of good SLM practices on the remaining challenges.

7.4 Why be interested in SLM?

The body of this book has been about the technological dimension of SLM. In this final chapter, however, we pay attention to broader concerns in business science, including the changing global marketplace and a firm's organizational/management structure.

Chapter 1 introduced SLM with a discussion of reasons why a business would be interested in SLM. It is fitting to end the book by listing those reasons again. It is likely that they will take on new meaning now.

▶ A enterprise infrastructure is a commodity for today's global business environment. In short, businesses depend on the unique services that are supported by the infrastructure. Thus, it is important for those services to be identified, understood, monitored, and controlled in accordance to some rational procedure.

▶ Many businesses have a large investment in their networks, systems, and applications. That investment is sometimes called the total cost of ownership (TCO) regarding the enterprise. Most businesses, however, have a hard time understanding the extent to which the enterprise contributes to business profit. But if one understands the services provided by the enterprise and the relation between profit and services (i.e., total benefits), then we can calculate the rate of return on investment (ROI). The usual equation for ROI is:

$$ROI = (total\ benefits - TCO)\ /\ total\ initial\ investment$$

SLM methods may help a business unpack the equation to see the utility of expenditures on enterprise components and management tools.

▶ A trend in business is to departmentalize. One such department is the IT department. The goal of the IT department is to establish and maintain the enterprise infrastructure for the business. Thus, the IT department must understand the service requirements of other departments in the business, for example, the sales, marketing, and research departments.

▶ Some businesses elect to outsource their IT requirements, in which case the IT department is under contract with the business. Other businesses elect to develop their IT department internally. In either case, expectations of services are set and the deliverance of services is monitored, and that calls for some rational procedure by which to do so.

▶ A contributory measure of a business's operational efficiency is the extent to which crucial services provided by the enterprise infrastructure are met. In addition, a good SLM program would identify weaknesses in service provisions, and a better SLM program would suggest ways to rectify the weaknesses.

▶ Finally, a contributory index into a business's value or prospective value is the extent to which the business is operating efficiently. An objective report on a business's service requirements and an evaluation of service provisions provide input into the overall value of the business and thus may influence potential investments into the company.

Summary

This chapter looked at SLM from a business and a human perspective. We discussed roles played by ISs in modern businesses, where an IS is broadly defined as a set of interrelated components that collect, process, store, and disseminate information for purposes of decision making, coordination, control, analysis, and production of commodities in an organization. We described the relation between ISs and SLM. Because ISs depend on the enterprise infrastructure (including network devices, computer systems, and applications), the SLM dimension becomes a crucial aspect of business success. The challenges faced by today's business include the SLM challenge. Finally, we considered reasons why a business would be interested in SLM.

Exercises and discussion questions

We repeat two instructive exercises from Chapter 3. We have covered a lot of technical material since Chapter 3, but, as we have seen in this chapter, technology is only part of the story. The other parts of business concern organization, management, and adaptability to a global marketplace.

For these two questions, pick an appropriate value of X, for example, banking, finance, government, health care, education, electronic commerce, or others.

1. Imagine that you are an SLM consultant and you have scheduled a one-hour meeting with a business executive of X organization. Outline an agenda for this first meeting.

2. Imagine you are a business executive of X organization. Put all SLM thoughts out of your mind. Pretend you know nothing about SLM. Now, an SLM consultant has managed to arrange a one-hour meeting with you, but you do not expect much to come of it. What could the consultant say to stimulate and pique your interests?

Further studies

Laudon and Laudon's textbook *Management Information Systems: New Approaches to Organization and Technology* influenced this chapter.

There is an abundance of literature on business science and ISs. The reader motivated in that direction will have no problem finding other relevant literature. It is unlikely that one will find discussions of SLM in current business literature, although we should expect to see such discussions in 2000+ literature.

For a good starter paper on current research for Internet2 and the challenges and issues, see the paper by Collins et al., "Data Express: Gigabit Junction With the Next Generation Internet." The paper is general, but technical. Also, visit www.internet2.edu. Much literature anticipates the general services that Internet2 will support and how those services are expected to contribute to business and the quality of life (e.g., distance learning, telemedicine, and telecommuting). For starters, see the *IEEE Communications Magazine* special issue on communications and telelearning.

Select bibliography

Collins, J., et al. "Data Express: Gigabit Junction With the Next Generation Internet." *IEEE Spectrum,* Feb. 1999.

IEEE Communications Magazine, special issue on Communications and Tele-Learning. Vol. 37, No. 3, Mar. 1999.

Laudon, K., and J. Laudon. *Management Information Systems: New Approaches to Organization and Technology* (5th ed). Englewood, Cliffs, NJ: Prentice Hall. 1998.

Epilogue

Back in Chapter 3, I quoted a colleague in Brazil who summed up the current state of affairs in SLM nicely.

> One of my intentions with the small business is dealing with the problem of SLM management methods. We wish to establish some kind of neutral methodology that can be applied to small/medium networks in order to have a better-managed environment. We know that larger networks deserve much attention; yet companies that have a well-established methodology sometimes run into trouble when deploying their products and techniques, particularly when it comes to integrating their products with products supplied by others and home-developed products. However, I remain optimistic even though I often have to adopt someone else's particular methodology when I deploy their products.

I noted that my colleague in Brazil not only was looking for a vendor-neutral methodology but was also a reviewer of my manuscript as the chapters unfolded. I suggested that by the end of the book we could perhaps revisit his predicament to see what progress had been made.

Well, the end of the book is here and this is what he had to say:

> After reading the complete book I can say that I now have a much better viewpoint from which I can evaluate all the processes that are involved in network management and its integration within a comprehensive IS/IT management effort. This viewpoint is needed

to put all things together, from IT systems development, to management system deployment, to meetings with the customers about their required services. I would recommend your book to anyone who wants to know more not just about SLM, but also about enterprise management in general.

By the way, when I wrote you about a "neutral" methodology, I couldn't wish for a better treatment than the one you have provided. The book precedes my thoughts and will give as much enlightenment to the reader as he/she is able to extract.

Congratulations on your work. I look forward to us meeting some day (real, instead of virtual).

So far, then, things look promising for my book. I look forward to hearing from other readers regarding the utility of the book and ways to improve on the concepts and methods described in it.

To conclude, let me take a moment to mention some work currently in progress. I am participating in an experiment to manage a GigaPOP in North Carolina that resides on the Next Generation Internet (also known as Internet2). Part of the experiment is to understand, define, and manage the unique services that the GigaPOP will provide. The challenges and results are too premature to include in this book, but I hope we will see a paper about it in the early 2000s.

List of acronyms and abbreviations

ACA   alarm correlation agent

AI   artificial intelligence

AMS   applications management system

AN   alarm notifier (Spectrum)

API   application programming interface

ASM   Adaptive System Management

BP   business process

BPMS   business process management system

CBR   case-based reasoning

CIM   Common Information Model

CLI   command line interface

CMIP   common management information protocol

CMU   Carnegie Mellon University

CORBA   common object request broker architecture

CPU   central processing unit

CRC   class-responsibility-collaboration

CS   computer system

CSA   computer system agent

DAI   distributed artificial intelligence

EC   electronic commerce

ECA   event correlation agent

EMS   enterprise management system

EMS   enterprise management system

e-mail   electronic mail

FCAPS   fault, configuration, accounting, performance, and security

FSN   full-service network

GUI   graphical user interface

GW   GlaxoWellcome

IA   infrastructure agent

IC   intercontinental communication

IDL   interface definition language

ILP   inductive logic programming

ISINM   International Symposium on Integrated Network Management

IT   information technology

L   link

LAN   local area network

MAN   metropolitan area network

MBR   model-based reasoning

MIB   management information base

MIT   Massachusetts Institute of Technology

MIT   management information tree

NC   NerveCenter

NMS   network management system

NSM   Network Security Manager

OMG   Object Modeling Group

OMS   off-line management system

OO   object oriented

OOA   object-oriented analysis

QA   quality assurance

QoS   quality of service

RBR   rule-based reasoning

ROI   return on investment

RT   response time

RTA   response time agent

S   server

SA   security agent

SDA   software distribution agent

SE   software engineering

SG   SpectroGRAPH

SLA   service-level agreement

SLM   service-level management

SLR   service-level report

SMS   systems management system

SNMP   simple network management protocol

SP   service parameter

SPT   spare part tracking

SS   SpectroSERVER

STG   state transition graph

TA   traffic agent

TCO   total cost of ownership

TINA   telecommunications information networking architecture

TMN   telecommunications management network

TMS   traffice management system

TTA   trouble-ticketing agent

telco   telephone company

U   user

UPS   uninterrupted power source

VoD   video on demand

WAN   wide area network

WM   working memory

About the author

Lundy Lewis is Director of Research at Cabletron Systems. He holds several patents in enterprise management and serves on the architectural board for the Spectrum Enterprise Management System. Dr. Lewis publishes in the professional literature and frequently gives presentations and tutorials at professional conferences. His first book, *Managing Computer Networks: A Case-Based Reasoning Approach*, was published by Artech House in 1995.

Dr. Lewis is an adjunct professor at the University of New Hampshire and New Hampshire College, where he teaches graduate-level courses in artificial intelligence, computer information systems, object-oriented methodology, and software engineering. He has taught courses in computer science and philosophy at Rensselaer Polytechnic Institute at Hartford, State University of New York at Binghamton, and the University of Georgia.

Dr. Lewis received a Ph.D. in philosophy from the University of Georgia, an M.S. in computer science from Rensselaer Polytechnic Institute, and a B.S. in mathematics and a B.A. in philosophy from the University of South Carolina. He is a member of IEEE, ACM, and AAAI.

Index

Adaptation, 165
  by substitution, 186
  critic-based, 186
  MBR, 176
  parameterized, 185
  STG, 179
Adaptive System Management
    (ASM) tool, 223
  propositional experiments, 230
  Top N algorithm, 229
Administration, 29
Agents
  alarm correlation, 148, 169
  application, 25, 38, 208
  autonomy, 124
  combination of, 26
  communication, 124–25
  defined, 37
  device, 25, 207
  enterprise, 26, 38, 87, 131, 208
  event correlation (ECA), 161, 162
  network, 37
  RBR, 169
  real-time, 208
  resources, 51, 124
  selection, 25–26, 207–10

  session, 51
  software distribution (SDA),
    189–90
  special-purpose, 25–26, 38, 208
  system, 25, 37, 207–8
  traffic, 25, 37, 87, 207
  type, 125
Alarm correlation
  across multiple domains, 147
  agent (ACA), 148, 169
  interdomain, 147
AlarmNotifier, 91
Alarm rollup, 208
Alarms
  component, 85
  defined, 80
  notification methods, 80
  objects in SLM domain, 82
  service, 85
Analysis model, 83–85
  control objects, 83
  defined, 83
  entity objects, 83
  interface objects, 83
  for View SLR use case, 84
  See also Use case methodology

Application agents, 25, 38, 208
Application response time (ART),
 189
Applications management systems
 (AMSs), 8
 examples, 8
 products, 9
 tasks, 8
 vendors, 8, 9
Architecture, 20–22, 111–54
 conceptual, 112, 115
 control, 21
 defined, 112–13
 inferencing, 21
 information, 279
 interpretation, 20
 multiloop, 128–30
 physical, 112–13, 116
 possibilities, 21
 single-loop, 127–28
 SLM evaluation with respect to,
  139–42
 strategic, 113
 subsumption, 130–32
 types, 20
Assembly line robot analogy,
 118–19
Asset management, 253, 254
Autodiscovery mechanisms, 209
Automated fault management,
 255–56
 defined, 255
 requirements, 256
Availability, 245

Backtracking, 99, 209
Behavior Modeling, 227
Blackboard system, 124
Business processes (BPs)
 defined, xi, 19, 36, 275
 definition of, 19
 elements of, 10
 inferring, 276

management system (BPMS), 10
 modeling, 150–51
 representation problem, 27–28
Business trends, 277–80
 information flow, 277–78
 unit organization, 277
 See also Modern business

Cabletron Spectrum, 87, 142
 alarm log, 269
 alarm notifier (AN), 148
 autodiscovery mechanism, 209
 C++ API, 212
 command line interface (CLI),
  148, 212
 computer systems management,
  210
 Data Warehouse, 87
 deployment, 145
 distributed client/server architecture, 143
 distributed version installation, 145
 event log, 268
 ICS screen shot, 264
 MBR approach, 173–76
 NerveCenter integration, 148–49
 NetCool integration, 148
 port status view, 266, 267
 SLA invocation, 265
 SpectroGRAPHs, 143
 SpectroRx, 150, 186–87
 SpectroSERVERs, 143
 SpectroWatch, 176
Cabletron Systems and AT&T case
  study, 44–46
 defined, 44
 possibilities, 45
 SLM framework, 44–45
Case-based reasoning (CBR),
  183–88
 architecture, 184
 CBR Express, 186
 challenges, 184, 185

defined, 183–84
generic systems, 186–87
goals, 183
SpectroRx, 186–87
summary, 187–88
See also Reasoning paradigms
Case studies
Cabletron Systems and AT&T,
44–46
Decysis, 102–6
Deutche Telekom, 142–51
GlaxoWellcome (GW), 16–18
KLM Airlines, 218–32
Windward Consulting Group,
246–66
Catalog Site, 243
CBR Express, 186
Change management, 253–54
Class cards, 94–97
defined, 94
generic, 94, 95
illustrated, 95, 96
stack, 94
for trouble-ticketing system, 96
See also CRC methodology
Classes, 92–93
alarm object, 93
defined, 92
hierarchy, 93
responsibilities, 93
signature, 94, 95
See also CRC methodology
Class-responsibility-collaboration.
See CRC methodology
Codebook approach, 179–83
codebooks, 180–82
coding, 180
correlation matrix, 179–80, 181,
182–83
See also Reasoning paradigms
Codebooks, 180–82
defined, 180
first, 180–81

multiple, choosing from, 182
second, 181
Code model, 89
Coding, 180
Collaboration, 94
Common Information Model
(CIM), 211
Common management information protocol (CMIP), 52
Common object request broker
architecture (CORBA),
52–53
acceptance, 53
defined, 52
standard, 52
Communication, 119
Complexity problem, 28, 215–18
Component parameters
defined, 36–37
service parameters vs., 41
Components
categories, 46
reuse, 90–91
in software system, 90
Component-to-service mapping,
24–25, 196–206
defined, 196
example, 196–97
fuzzy logic, 198–206
parameter, 37
Compositional rule of inference,
206
Computational overhead, 165, 179
Conceptual architecture, 112
EC, 261
enterprise management, 138
illustrated, 115
See also Architecture
Conceptual graph, 39–44
concepts, 39, 40
defined, 35
final increment, 43
first increment, 40

notation, 39
relations, 39, 40
second increment, 41, 42
Configuration management
defined, 253
procedure application, 253
requirements, 254
See also Electronic commerce (EC)
Continuity product, 87
Contracts
defined, 93
signature, 94
Control objects, 83
Correlation matrix, 179–80, 181,
    182–83
CRC methodology, 76, 92–98
appeal of, 92
class cards, 94–97
classes, 92–93
collaboration, 94
contracts, 93
features, 97
with OO language, 92
responsibilities, 93–94
scenario, 92
subsystems, 97
See also SE methodologies
Critic-based adaptation, 186

Dataconference, 274
Data integrity, 194
Data marts
data warehouse combined
    scheme, 136
defined, 135
functionally distributed scheme,
    135
Data mining, 219–24
amenable tasks, 231–32
in enterprise management, 231–32
motivation, 219
parameter vectors, 220
Data scrubbing, 134

Data warehouses, 134–35
Data warehousing, 133–38
application to enterprise manage-
    ment, 133
data mart, 135
data scrubbing, 134
data warehouse, 134–35
illustrated, 134, 136
methods, 133
operational data, 133–34
Decision tree algorithms, 221
Decision trees
multiparameter (misbehaving sub-
    net), 227
multiparameter (SPT), 228
multiparameter, 225
produced by Tilde (SPT), 231
structure of, 222
two-parameter, 225
Decysis case study, 102–6
defined, 103
insights, 104–5
methodology steps, 104
SLA review cycle, 105
SLM methodology, 103–4
Deliberative behavior, 128
Deployment phase
defined, 68
production/baseline step, 73
SLR review step, 73
SLR/SLA review step, 73
step outline, 68
See also SLM methodology
Design, unit testing, and integra-
    tion testing phase
agent integration step, 72–73
component inventory step, 70–71
component parameter mapping
    step, 72
defined, 68
service/component correlation
    step, 71
step outline, 68

See also SLM methodology
Design model, 86–89
  defined, 86
  for View SLR use case, 88
  See also Use case methodology
Deutche Telekom case study,
      142–51
  BP modeling, 150–51
  EMS, 142
  fault management, 146
  multiple layers of abstraction,
      146–50
  service agreement, 146
  Spectrum, 142–45
Device agents, 25, 207
Digex, 243
Distributed artificial intelligence
      (DAI), 120–26
  defined, 120–21
  dimensions of multi-agent
      systems, 123–25
  enterprise management problem,
      121
  general rules, 125–26
  problem characteristics, 121
  problem dimensions, 122
Distributed processing, 49, 50
Domain model, 80–82
  with boundary, 81
  defined, 80
  illustrated, 81
  See also Use case methodology

Effectors, 119
Electronic commerce (EC), 28–29,
      241–70
  asset management, 253, 254
  automated fault management,
      256–57
  bundled services, 245
  change management, 253–54
  conceptual EMS architecture, 261

configuration management, 253,
      254
  defined, 242–43
  EMS console, 258–59
  EMS requirements, 249–60
  event management, 251–52
  growth, 243
  market, 242, 259
  monitoring, 250–51
  organizational structure, 248–49
  physical architecture, 263
  problem management, 255, 256
  products, 245
  reporting, 252–53
  security management, 257
  server farms, 259
  service parameters, 245
  SLA management, 257–58
  SLAs, 245
  software distribution, 254–55
  supplier burdens, 243–44
  trend and performance analysis,
      256–57
End-to-end SLM, 27, 212
  defined, 26
  selective SLM vs., 27, 213
  See also Selective SLM
Enterprise agents
  candidates, 87
  defined, 26, 38, 208
  event correlation, 131–32
  See also Agents
Enterprise management
  conceptual architecture, 138
  data mining in, 231–32
  data warehousing application to,
      133
  off-line, 137
  perspective, 120
  real-time, 137
  solutions, 114
  tasks, 113
  truths, 113–14

Enterprise management systems
        (EMSs), 9
    console, 258–59, 262
    Deutche Telekom, 142
    for EC business, 249–60
    GlaxoWellcome, 16
    integration, 114
    intelligence in, 118
    vendors, 10
Enterprise networks
    as commodities, 14
    complexity of, 46
    defined, xi, 36, 275
    simple, illustrated, 6
    trucking operation similarities, 5
    See also Enterprise management
Entity objects, 83
ETEWatch, 210
Event correlation, 80–82, 86,
        131–32, 158–90
    challenge assumptions, 160–61
    codebook approach to, 179–83
    defined, 23, 208
    distributed, 188–90
    distributing, over multiple agents,
        170
    example, 159–65
    intradomain, 146
    model-based reasoning approach,
        171–76
    rule-based reasoning approach,
        165–71
    state-transition graph (STG) ap-
        proach, 176–79
Event correlation agent (ECA), 161
    central bucket and, 162
    human, 162–64
    reasoning paradigms, 164
    task, 164
Event management, 251–52
    defined, 251
    prominence of, 260
    requirements, 252

See also Electronic commerce (EC)
Event-to-alarm mapping, 182
Exercises/discussion questions
    concepts and definitions, 61
    modern business, 282
    SLM and electronic commerce,
        267–68
    SLM architecture, 152–53
    SLM introduction, 30–31
    SLM methodology, 106–7
    special topics, 234

Fault management, 146
    automated, 255–56
    multidomain, 147
File Access Influence, 224
Football team analogy, 27
Full-service network (FSN), 49–50
Further studies
    concepts and definitions, 62–63
    modern business, 282–83
    SLM and electronic commerce,
        268–70
    SLM architecture, 153–54
    SLM introduction, 31–32
    SLM methodology, 107–8
    special topics, 234–35
Fuzzy inference engine, 206
Fuzzy logic, 198–206
    engineering framework, 204
    "heavy" concept in, 202–3
    "light" concept in, 203
    network parameters in, 202
    "ok" concept in, 203
    operation, for service manage-
        ment, 205
    rule examples, 205

GlaxoWellcome (GW) case study,
        16–18
    EMS, 16
    enterprise infrastructure develop-
        ment, 16

IC response measurement, 18
management architecture, 17
summary, 18
Globalization challenge, 279
GTE Internetworking, 243, 262

Happy-medium approach
    defined, 192
    periodic database transaction
        execution, 193
    See also Semantic disparity problem
HP OpenView, 87
    autodiscovery mechanism, 209
    computer systems management,
        210
    event monitoring, 178
    InCharge integration, 182
    NerveCenter integration, 178
Human intelligence, 158–59
Human sympathy, 273

IBM NetView, InCharge integra-
        tion, 182
InCharge, 176, 182
    event-modeling language, 182
    integration, 182
    networking classes, 182
Inductive logic programming (ILP)
        algorithms, 223, 230
Information architecture, 279
Information systems
    1970s view, 272–73
    2000+ view of, 273–75
    as combined organizational-
        management-technological
        solution, 277
    defined, 272, 275
    development/maintenance of, 277
    dimensions, 277
    introduction of, 276
    investment challenge, 280
    modern business and, 272–75
    modern day, 273

QA engineer's use of, 273
SLM connection, 275–77
using, effectively, 277
Information technology (IT)
        department
    development, 15
    goal of, 15
    responsibilities, 277
Integrated management, 46–53
    CMIP, 52
    configuration illustration, 54
    CORBA, 52–53
    discussion, 53
    SNMP, 51–52
    TINA architecture, 49–51
    TMN model, 46–49
Integration, 26
    approaches, 212
    architecture, 262
    problem, 210–12
    projects, 211
Integrity, 245
Intercontinental e-mail, 190–91
Interdomain alarm correlation, 147
Interface definition language
        (IDL), 52
Interface objects, 83
Internet2, 175
Intradomain event correlation, 146

Jitter, 194
Jyra agent, 210

KLM Airlines case study, 218–32
    data mining concepts, terms, algo-
        rithms, 219–24
    data mining motivation, 219
    defined, 218
    lab experiments, 224–27
    monitoring agents, 228
    SPT multiparameter decision tree,
        228
    SPT service, 228

Knowledge acquisition, 164
  codebook approach, 182
  STG approach, 179
Knowledge-level systems, 278
Knowledge representation, 164,
    179

Learning, 165
  MBR, 175
  STG, 179
  See also Reasoning paradigms

Management information bases
    (MIBs), 51
Management-level systems, 278
Methodologies. *See* SE method-
    ologies; SLM methodology
Misbehaving Subnet, 226
Model-based reasoning (MBR),
    171–76
  adaptability, 175
  for event correlation, 173
  example, 171–72
  learning, 175
  Spectrum, 173–76
  success, 176
  See also Reasoning paradigms
Modern business, 29
  1970s view of, 272–73
  2000+ view of, 273–75
  features for SLM discussion, 276
  information systems and, 272–75
  trends, 277–80
Modus Ponens inference rule, 166
Monitoring, 250–51
  application requirements, 251
  computer system requirements,
    250
  end-to-end requirements, 251
  facilities requirements, 251
  network requirements, 250–51
  See also Electronic commerce (EC)
Monitoring agents. *See* Agents

Multi-agent systems, 123–25
  adaptability, 123–24
  agent autonomy, 124
  agent communication, 124–25
  agent resources, 124
  agent type, 125
  grain, 123
  model, 123
  scale, 123
Multiloop architecture, 128–30
  control loops, 128–29
  defined, 128
  illustrated, 129
  single-loop architecture vs., 130
  See also Robot intelligence
Multimedia services, 50

NerveCenter, 148–49, 176
  OpenView integration, 178
  RBR method, 179
NetCool, 148, 169–70
Network agents, 37
Network management systems
    (NMSs), 6
  event correlation, 7
  SMSs vs., 7
  vendors, 8
  views, 7
Network Security Manager (NSM),
    170

Object Modeling Group (OMG), 53
  QoS agreements, 58
  QoS categories, 59
  QoS characteristics and parame-
    ters, 57–58
  QoS management, 59–60
  QoS management functions, 58–59
  security, 55–56
  service, 60
  static vs. adaptive QoS, 56–57
  system performance vs. system
    function, 54–55

user vs. engineering view
  points, 55
Object-oriented analysis (OOA),
  172
Ockham's Razor, 39
Off-line enterprise management,
  137
Off-line management system
  (OMS), 138
Operational data, 133–34
Operation-level systems, 277–78
Organization, this book, 18–29

Parameterized adaptation, 185
Physical architecture, 112–13
  application-centric, 117
  EC, 263
  illustrated, 116, 117
  network-centric, 116
  See also Architecture
Policies and business rules, 119
Private responsibilities, 93–94
Problem management
  defined, 255
  requirements, 256
  See also Electronic commerce
  (EC)
Progol, 223, 230
Propositional representations, 220
Push and pull technology, 124

Quality of service (QoS)
  adaptive, 56
  agreements, 58
  categories, 59
  characteristics and parameters,
  57–58
  management, 53–60, 59–60
  management functions, 58–59
  standards, 54
  static, 56
  system performance, 54–55
  viewpoints, 55

Quantified representations, 220

Real-time agents, 208
Real-time enterprise management,
  137
Reasoning
  agent item, 119
  algorithm, 166
Reasoning paradigms
  adaptation, 165
  case-based, 183–88
  codebook, 179–83
  computational overhead, 165
  dimensions, 164–65
  knowledge acquisition, 164
  knowledge representation, 164
  learning, 165
  list of, 164
  model-based, 171–76
  rule-based, 165–71
  scalability, 165
  state-transition graph, 176–79
Reflexive behavior, 128
Reporting, 252–53
Representation problem, 27–28,
  213–15
  analogy, 27, 213–14
  BP, 27–28
Requirements and analysis phase
  consumer BP understanding step,
  69
  defined, 67
  enterprise-related services understanding step, 69
  service parameters/levels understanding step, 70
  step outline, 68
  See also SLM methodology
Resource agent, 51
Response time, 194
  application (RTA), 189
  EC, 245
  service parameter, 220, 224

Responsibilities, 93–94
    defined, 93
    private, 93–94
    signature, 94
    See also CRC methodology
Return on investment (ROI),
        14–15, 280–81
RMON II+ agent, 87, 89
Robot intelligence, 126–33
    architecture debate, 132–33
    development approaches, 126–27
    multiloop architecture, 128–30
    single-loop architecture, 127–28
    subsumption architecture, 130–32
    summary, 132–33
Rule-based reasoning (RBR),
        165–71
    agent, 169
    challenges, 168
    elements, 165
    in enterprise management, 169
    lean semantics, 201
    for load notices, 201
    reasoning algorithm, 166
    rule base, 165–66
    structure, 166
    system construction, 199–200
    use of, 168
    variations, 199
    working memory, 165
    See also Reasoning paradigms
Rule induction algorithms, 221

Scalability problem, 165, 179
Scaling, 26–27, 212–13
SE
    defined, 74
    projects, 74, 75, 90
    unsuccessful, program, 75
    See also SE methodologies
Security
    EC, 245
    management, 257

OMG, 55
SLM, 56
Selective SLM, 27, 212–13
    defined, 27
    end-to-end SLM vs., 27, 213
    See also End-to-end SLM
Semantic disparity problem,
        13–14, 23–24, 190–96
    approaches, 13, 192
    choosing approach for, 14
    defined, 13, 190
    happy-medium approach, 192
    service concept and, 191
    techno-centric approach, 192
    user-centric approach, 192, 206
SE methodologies, 65–66, 74–98
    CRC, 76, 92–98
    goal, 76
    use case, 76, 77–92
Sensors, 119
Service availability, 195
Service level agreements (SLAs)
    defined, 37
    development challenge, 104
    EC, 245
    elements, 41, 43–44
    implementation, 104
    initial setup of, 105
    invocation, 265
    management, 257–58
    performance metrics, 245
    review cycle, 105
    sample form, 247
    templates, 44
Service level management. *See* SLM
Service level reports (SLRs)
    defined, 37
    elements, 41–42
    View use case, 84, 88
Service levels, 37
Service parameters, 192
    capacity, 45
    component parameters vs., 41

defined, 36
EC, 245
list of, 193
response time, 220, 224
Services, 192
  bundled, 245
  defined, xi, 9, 19, 36, 275–76
  EC, 245
  in enterprise space, 9
  multimedia, 50
  OMG, 60
  trouble-ticketing, 45
Session agent, 51
Simple network management protocol (SNMP), 51–52
  components, 51
  defined, 51
  limitations, 52
Single-loop architecture, 127–28
  defined, 127
  illustrated, 127
  multiloop architecture vs., 130
  sensor data abstraction, 128
  See also Robot intelligence
SLM
  administration of, 29
  agents, 25–26
  architecture, 20–22, 111–54
  case study, 16–18
  challenge, 280
  conceptual graph, 35, 39–44
  course exam questions, 140–42
  crux of, 12–14
  current state-of-affairs in, 103
  defined, 2–5, 38
  design/implementation problems, 22–28
  electronic commerce and, 28–29, 241–70
  end-to-end, 26, 27, 212, 213
  evolution timeline, 11
  evolution toward, 5–12
  framework, 36, 44–45, 56

interest in, 280–81
introduction to, 1–32
IS connection, 275–77
practical business need for, 14–15
process, 29
proposal evaluation with respect to architecture, 139–42
reasons for pursuing, 14–15
selective, 27, 212–13
summation, 285
truck operation analogy, 2–5
use case model, 78–79
SLM methodology, 19–20, 65–108
  backtracking, 99
  Decysis case study, 102–6
  deployment, 67
  design, unit testing, and integration testing, 67
  development guide, 66
  essential, 68
  phases, 67
  requirements and analysis, 67
  sequential steps, 209
  "starting at the bottom" strategy, 101–2
  "starting at the top and bottom" strategy, 102
  "starting in the middle" strategy, 100–101
  summary, 106
  variations, 99–102
  "version 1 space, version 2 space" strategy, 100
Software distribution, 254–55
  agent (SDA), 189–90
  defined, 255
  requirements, 255
  See also Electronic commerce (EC)
Special-purpose agents, 25–26, 38, 208
SpectroGRAPHs, 143
SpectroRx, 186, 264
  anecdote, 186–87

defined, 150
SpectroSERVERs, 143
  central master, 145
  communication by intermediary
    agents, 144
  deployment, 146
  maximum ratio, 144
  See also Cabletron Spectrum
SpectroWatch, 276
Spectrum. *See* Cabletron Spectrum
State-transition graph (STG)
  adaptability, 179
  approach, 176–79
  computational overhead, 179
  concepts, 176
  enterprise behavior, 227
  instrument, 178
  knowledge acquisition, 179
  knowledge representations, 179
  learning, 179
  sample, 177
  scalability, 179
  for weekly account aggregation,
    216
  See also Reasoning paradigms
StopWatch, 210
Strategic architecture, 113
Strategic business challenge, 279
Strategic-level systems, 278
Subsumption architecture, 130–32
  arguments against, 132
  behaviors, 130
  defined, 130
  demonstration, 131
  illustrated, 131
  sensor data, 130
  See also Robot intelligence
Subsystems
  defined, 97
  illustrated, 98
  See also CRC methodology
System agents, 25, 37, 207–8
System performance, 54–55

Systems management systems
  (SMSs), 7
  NMSs vs., 7
  tasks, 7
  vendors, 8

Techno-centric approach
  to availability, 194
  data, 196
  defined, 192
  See also Semantic disparity problem
Telecommunications information
  networking architecture.
  See TINA
Telecommunications management
  network (TMN), 11
  business/enterprise management
    layer, 47
  characteristics, 48
  defined, 46
  GUI interoperability standards, 48
  management layers, 12, 47
  management tasks, 47–48
  model, 46–49
  network element layer, 48
  network element management
    layer, 47
  network management layer, 47
  service management layer, 11, 47
  TINA comparison, 49
Teleconference, 274
Test model, 89
Thrashing, 99
Tilde, 224, 230, 231
TINA, 49–51
  defined, 49
  distributed processing, 49, 50
  management layers, 49
  TMN layer comparison, 49
Top N algorithms, 221, 229
Total cost of ownership (TCO), 14,
  280
Traffic agents

candidate, 87
defined, 25, 37, 207
Traffic management systems
    (TMSs), 7, 8
Trend/performance analysis,
    256–57
Trouble ticketing
    class card for, 96
    service, 45
Truck operation analogy, 2–5
    enterprise network similarities, 5
    functions/personnel, 2–3
    illustrated, 2, 4
    main points, 4–5

Usability challenge, 280
Use case methodology, 76, 77–92
    analysis model, 83–85
    code model, 89
    considerations, 90–91
    design model, 86–89
    domain model, 80–82
    goal, 91
    illustrated, 77

test model, 89
use case model, 78–79, 91
See also SE methodologies
Use case model, 78–79, 91
    defined, 78
    SLM, 78–79
User-centric approach, 206
    data, 196
    defined, 192
    transaction type testing, 193
    See also Semantic disparity problem

Videoconference, 274
View SLR use case, 84, 88
Virtual classroom, 274
Virtual local area networks
    (VLANs), 71
Vital Agent, 210

Web server farms, 28–29
Windward Consulting Group case
    study, 242, 246–66
Working memory (WM), 199, 200

Recent Titles in the Artech House Telecommunications Library

Vinton G. Cerf, Senior Series Editor

Access Networks: Technology and V5 Interfacing, Alex Gillespie

Achieving Global Information Networking, Eve L. Varma, Thierry Stephant, et al.

Advanced High-Frequency Radio Communications, Eric E. Johnson, Robert I. Desourdis, Jr., et al.

Advances in Telecommunications Networks, William S. Lee and Derrick C. Brown

Advances in Transport Network Technologies: Photonics Networks, ATM, and SDH, Ken-ichi Sato

Asynchronous Transfer Mode Networks: Performance Issues, Second Edition, Raif O. Onvural

ATM Switches, Edwin R. Coover

ATM Switching Systems, Thomas M. Chen and Stephen S. Liu

Broadband Network Analysis and Design, Daniel Minoli

Broadband Networking: ATM, SDH, and SONET, Mike Sexton and Andy Reid

Broadband Telecommunications Technology, Second Edition, Byeong Lee, Minho Kang, and Jonghee Lee

Client/Server Computing: Architecture, Applications, and Distributed Systems Management, Bruce Elbert and Bobby Martyna

Communication and Computing for Distributed Multimedia Systems, Guojun Lu

Communications Technology Guide for Business, Richard Downey, Seán Boland, and Phillip Walsh

Community Networks: Lessons from Blacksburg, Virginia, Second Edition, Andrew Cohill and Andrea Kavanaugh, editors

Computer Mediated Communications: Multimedia Applications,
Rob Walters

Computer Telephony Integration, Second Edition, Rob Walters

Convolutional Coding: Fundamentals and Applications,
Charles Lee

Desktop Encyclopedia of the Internet, Nathan J. Muller

*Distributed Multimedia Through Broadband Communications
Services,* Daniel Minoli and Robert Keinath

Electronic Mail, Jacob Palme

*Enterprise Networking: Fractional T1 to SONET, Frame Relay to
BISDN,* Daniel Minoli

FAX: Facsimile Technology and Systems, Third Edition, Kenneth R.
McConnell, Dennis Bodson, and Stephen Urban

Guide to ATM Systems and Technology, Mohammad A. Rahman

Guide to Telecommunications Transmission Systems,
Anton A. Huurdeman

A Guide to the TCP/IP Protocol Suite, Floyd Wilder

Information Superhighways Revisited: The Economics of Multimedia,
Bruce Egan

International Telecommunications Management, Bruce R. Elbert

Internet E-mail: Protocols, Standards, and Implementation,
Lawrence Hughes

Internetworking LANs: Operation, Design, and Management,
Robert Davidson and Nathan Muller

Introduction to Satellite Communication, Second Edition,
Bruce R. Elbert

Introduction to Telecommunications Network Engineering,
Tarmo Anttalainen

Introduction to Telephones and Telephone Systems, Third Edition,
A. Michael Noll

LAN, ATM, and LAN Emulation Technologies, Daniel Minoli and
Anthony Alles

The Law and Regulation of Telecommunications Carriers,
Henk Brands and Evan T. Leo

Marketing Telecommunications Services: New Approaches for a Changing Environment, Karen G. Strouse

Mutlimedia Communications Networks: Technologies and Services,
Mallikarjun Tatipamula and Bhumip Khashnabish, Editors

Networking Strategies for Information Technology, Bruce Elbert

Packet Switching Evolution from Narrowband to Broadband ISDN,
M. Smouts

Packet Video: Modeling and Signal Processing, Naohisa Ohta

Performance Evaluation of Communication Networks,
Gary N. Higginbottom

Practical Computer Network Security, Mike Hendry

Practical Multiservice LANs: ATM and RF Broadband,
Ernest O. Tunmann

Principles of Secure Communication Systems, Second Edition,
Don J. Torrieri

Principles of Signaling for Cell Relay and Frame Relay,
Daniel Minoli and George Dobrowski

Pulse Code Modulation Systems Design, William N. Waggener

Service Level Management for Enterprise Networks, Lundy Lewis

Signaling in ATM Networks, Raif O. Onvural and Rao Cherukuri

Smart Cards, José Manuel Otón and José Luis Zoreda

Smart Card Security and Applications, Mike Hendry

SNMP-Based ATM Network Management, Heng Pan

Successful Business Strategies Using Telecommunications Services,
Martin F. Bartholomew

Super-High-Definition Images: Beyond HDTV, Naohisa Ohta

Telecommunications Department Management, Robert A. Gable

Telecommunications Deregulation, James Shaw

Telemetry Systems Design, Frank Carden

Teletraffic Technologies in ATM Networks, Hiroshi Saito

Understanding Modern Telecommunications and the Information Superhighway, John G. Nellist and Elliott M. Gilbert

Understanding Networking Technology: Concepts, Terms, and Trends, Second Edition, Mark Norris

Understanding Token Ring: Protocols and Standards, James T. Carlo, Robert D. Love, Michael S. Siegel, and Kenneth T. Wilson

Videoconferencing and Videotelephony: Technology and Standards, Second Edition, Richard Schaphorst

Visual Telephony, Edward A. Daly and Kathleen J. Hansell

Winning Telco Customers Using Marketing Databases, Rob Mattison

World-Class Telecommunications Service Development, Ellen P. Ward

For further information on these and other Artech House titles, including previously considered out-of-print books now available through our In-Print-Forever® (IPF®) program, contact:

Artech House	Artech House
685 Canton Street	46 Gillingham Street
Norwood, MA 02062	London SW1V 1AH UK
Phone: 781-769-9750	Phone: +44 (0)20 7596-8750
Fax: 781-769-6334	Fax: +44 (0)20 7630-0166
e-mail: artech@artechhouse.com	e-mail: artech-uk@artechhouse.com

Find us on the World Wide Web at:
www.artechhouse.com